If the *mature* gills are white, p...
(*not* brown or black) and i...
the young mushroo...
is encased by a veil...
that forms a volva
(sack, scaly rings,
or collar) at base of
stalk *and/or* leaves flakes or patches on cap —

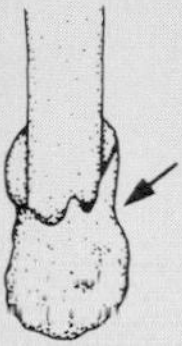

**Amanita
pp. 63-80**

If not as above

If there is no veil covering
the young gills, or if there
is only a thin veil that
quickly disappears and does
not form a ring on stalk

If the young gills
are covered by
a veil that usually
forms a ring on stalk

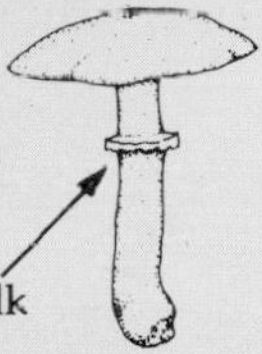

If the spores are chocolate-
brown and the *mature* gills
are deep chocolate-
brown (but white
or pink when
young) and free from stalk —

**Agaricus
pp. 100-123**

If the spores
and *mature* gills
are differently
colored or if
the gills are
attached to
the stalk

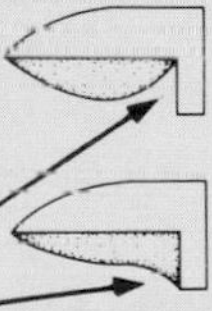

If the spores are white
(or in one case greenish)
and the *mature* gills are
white, yellow, or greenish —

**Miscellaneous
Light-spored
Gilled Mushrooms
with a Ring
pp. 81-99**

If the spores are rusty-
orange to brown or black
(and the *mature* gills are
variously colored but
not white) —

**Miscellaneous
Dark-spored
Gilled Mushrooms
pp. 124-151**

It is no dream!
Matsutake are growing
On the belly of the mountain.
—Shigetaka

All That the Rain Promises, and More...

A HIP POCKET GUIDE TO WESTERN MUSHROOMS

David Arora

Ten Speed Press
Berkeley

BioSystems Books
Santa Cruz

All rights reserved. Published in the United States by Ten Speed Press,
an imprint of the Crown Publishing Group, a division of Random
House, Inc., New York.
www.crownpublishing.com
www.tenspeed.com

Ten Speed Press and the Ten Speed Press colophon are registered
trademarks of Random House, Inc.

The haiku were translated by R.H. Blyth ("Mushrooms in
Japanese Verse," Transactions of the Asiatic Society of Japan, Vol.
XI, 1973).

Grateful acknowledgment is made to Vintage International,
Random House, for permission to reprint an excerpt from
Speak, Memory by Vladimir Nabokov.

*The author welcomes information and feedback from readers.
You can write to him in care of Ten Speed Press.*

Library of Congress Cataloging-In-Publication Data

Arora, David
    All that the rain promises, and more . . . :
    a hip pocket guide to western mushrooms

    Includes index.
    1. Mushrooms — West (U.S.) — Identification.
    I. Title

QK617.A688   1990
589.2′22′0978 — dc20   90-36171

ISBN-13: 978-0-89815-388-0

Printed in China

Typeset by Jonathan Peck Typographers Ltd. and Typola

32

First Edition

Preface

All That the Rain Promises, and More is a pocket-sized field guide to the most distinctive wild mushrooms of western North America. Stretching from the Rocky Mountains to the Pacific Coast and from the deserts of northern Mexico to Alaska, this vast and varied region harbors several thousand kinds of mushrooms. The approximately 200 depicted here are those most likely to pique the interest of the average wonderer and wanderer. A more comprehensive treatment isn't practical in a book this size. Witness my joy and pride, the comprehensive guide *Mushrooms Demystified*. It is eminently quotable but not notable for being totable. The only pocket it will fit in is a kangaroo's!

The design of this book is simple: each mushroom is illustrated with one or more color photographs accompanied by a concise, easy-to-understand list of identifying features plus information on edibility and habitat. In other words, it is a completely self-contained field guide for beginners. But it will also appeal to more experienced hunters as a field companion to *Mushrooms Demystified*, and is cross-referenced for that purpose. Even the jaded mushroom junkie whose cupboard runneth over with mushroom memorabilia will find something new: several species are depicted in color for the first time, and unusual uses of mushrooms are documented.

In a fungophobic (mushroom-loathing) society such as ours, it takes a certain boldness and curiosity to seek out mushrooms, and creativity to put them to good use. As a result, North American mushroom hunters are an unusually bold, curious, creative, even iconoclastic and bizarre bunch that have nourished and enhanced my life immeasurably. I have attempted to introduce some of them to readers by lacing this book with a generous helping of first-person accounts, novel cooking suggestions, stories, limericks, and photographs of them "caught in the act" of doing things with or to their favorite mushrooms. I am grateful to each and every one of them for their contributions and camaraderie. Few of us get to write books. Even fewer get to write books about mushrooms. As one who does both, sharing the insights and experiences of those who don't is a fitting way to record a small but significant part of our future history and cultural heritage.

One final note to readers unfamiliar with *Mushrooms Demystified*: I firmly believe in stressing the *fun* in *fun*gi. In leafing through these pages, you may wonder what all the "fanciful," "foolish," or (shudder) "extraneous" material is doing in a factual guide. After all, it is the practical, hands-on, how-to-identify information that makes this book useful and gives it substance. But I ask: is it any stranger or less desirable to sprinkle the facts with flakes of fancy than it is to liven up solemn, substantial fare like potatoes with something fancier and more flavorful, like wild mushrooms?

I say: thank goodness for the potatoes, but thank *God* for mushrooms!

—David Arora
Santa Cruz, California

A "mushroom-minded" individual: the author with candy cap cookies (p. 23) and mushroom toast (p. 31).

Contents

I used to think I needed the sun to have fun. Rain was an inconvenience, something to wait out, not wade in. The farmers needed it. I didn't. Rain meant I couldn't do things. It was the enemy of activity, the bane of beach barbecues, an imposition from above that didn't have the courtesy to call ahead.

Mushrooms changed all that. Now when it rains, I can't wait to get out, to plunge into that pristine, misty realm of glistening freshness and fleeting fragrance to see what new wonders the earth has to offer. The miracle of mushrooms is in their spontaneity and resilience. Springing from ground that looked so hard and bleak, they seem to embody all that we carry, and bury, inside us: secret passions and dormant dreams awaiting inspiration, instigation, and conditions that precipitate growth. Rain has become my catalyst, drawing me up, bringing me out.

I still savor the sun — who doesn't? Rain refreshes, sunshine caresses. But as I bask in the hazy glow of another lazy summer day, my life feels as empty as the sky above, and as surely as the shivering survivors of winter look forward to the spring, I find myself yearning for clouds returning, all that the rain promises, and more . . .

Introduction

Mushrooms are the reproductive structures or "fruit" of certain fungi. The most familiar kind of mushroom has a cap with *gills* (radiating blades) on its underside. Millions of microscopic reproductive units called *spores* are discharged from the gills and dispersed by air currents. Only a small percentage of spores land in a favorable environment, where they germinate to form new fungi.

Fungi do not manufacture their own food like plants. In this respect they are like animals: they must obtain food from outside sources. The part of the mushroom fungus that digests nutrients is an intricate web of fine threads collectively called the *mycelium* (plural: mycelia). The mycelium may live anywhere from a few days (in perishable substrates like dung) to several hundred years, periodically producing mushrooms when enough moisture is available.

Mushrooms, or more exactly the fungi that produce them, are a vital and omnipresent part of our environment. Despite their bad press, the overwhelming majority are beneficial. A few are *parasitic*, feeding on living organisms, usually trees. The rest are either saprophytic or mycorrhizal. *Saprophytic* fungi are nature's recyclers. They replenish the soil by breaking down complex organic matter (wood, dung, humus, etc.) into simpler, reusable compounds. *Mycorrhizal* fungi form a mutually beneficial relationship with the rootlets of plants in which nutrients are exchanged. They are critical to the health of our forests, as many trees will not grow without them. (Since some mycorrhizal mushrooms are associated with certain kinds of trees, make a habit of noting the different trees growing in the vicinity of any mushroom you wish to identify.)

Despite the many benefits and uses of mushrooms, most North Americans are markedly *fungophobic*, a trait inherited from the British. Fungophobia can be defined as the belief that mushrooms are actively hostile at worst and worthless at best. It is only in the last few years that large numbers of North Americans have begun to discover what the mushroom-loving peoples of Japan, China, Russia, and Europe have known for centuries: that these "forbidden fruit" are delicious and nutritious, vital and valuable, potent and beautiful, and that mushroom hunting is a challenging, enlightening, and uplifting activity.

How To Use This Book

Simple. Once you've collected a distinctive mushroom, consult the quick key to mushroom groups on the inside front and back covers, and go to the section of the book indicated. Flip through the pictures in that section until you find one that looks similar to your mushroom. Then carefully go through the numbered list of "Key Features" on that page, checking them off as you go so you don't inadvertently miss one. If your mushroom has *all* of the key features, then you have identified it! To verify your identification, check the details listed under "Other Features," "Where," etc.

If your mushroom does not agree with one of the "Key Features," assume it is different. First consult the section called "Note" at the bottom of the page for a possible explanation of the discrepancy or a listing of similar species, then continue searching the pictures for another likely candidate. If you can't find a photograph and set of key features that match your mushroom, there are two possible explanations.

The first is that your mushroom is a "freak" — an untypical example of an illustrated species, for instance one that has faded badly or lost its ring. By collecting *several* examples of each kind of mushroom you wish to identify, in different stages of development if possible, you are much more likely to gather at least a few typical, identifiable ones; then you can discard those that lack one or more of the key features unless you are experienced enough to recognize them. (Note: different kinds of mushrooms often mingle with one another; unless they are obviously the same, assume that they are different until you discover otherwise.)

The second, and more likely, explanation is that your mushroom is not in this book. After all, less than 10 percent of the known species from western North America are. You can return unidentified mushrooms to their place of origin, relegate them to the compost pile, or, if you are determined to know their identity, consult a more comprehensive guide such as *Mushrooms Demystified*, 2d ed. (1986).

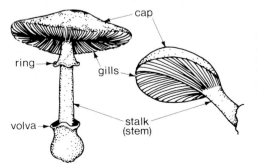

The parts of a gilled mushroom (see facing page for details).

Format and Terminology

Beginners are often frustrated by mushroom descriptions because they don't know how much importance to attach to each feature described. The odor or type of gill attachment may be pivotal to the recognition of one mushroom and completely irrelevant to the identification of another. In this book complete descriptions have been abandoned in favor of lists called "Key Features" and "Other Features." These and other aspects of the format are discussed below, along with ambiguous or unfamiliar terms.

A typical page includes a photograph of a mushroom, its common and scientific names, and the following headings:

Other Names (sometimes omitted). A brief list of the most widely used common (English) and scientific (Latin) synonyms that facilitates cross-reference with other mushroom books (many mushrooms have been classified in more than one way, and the common names are *not* standardized).

Key Features. A unique set of features that distinguishes the mushroom from others known to occur in western North America. This is the most important part of the text.

Other Features. Characteristics not critical to the identification of the pictured mushroom, but useful for verification should your mushroom have all of the key features.

With the following exceptions, the terms used in the above two sections are self-evident or are defined on the page where they are used.

A *veil* is a layer of protective tissue that typically extends from the stalk to the edge of the young cap. When the cap expands, the veil breaks, often forming a *ring* on the stalk. In some instances (the amanitas) the veil is called the *partial veil* to distinguish it from the *universal veil,* which completely encloses the young mushroom. Upon rupturing the universal veil often forms a *volva* (sack, cup, collar, or series of scaly rings) at the base of the stalk, and sometimes leaves flakes, warts, or a patch on the cap.

The size of a mushroom is a function of its age, the amount of moisture and nutrients available, and of course, its genetic makeup. Absolute size is not nearly as important as *average size range,* as follows:

> *large*: cap >6" broad; top of stalk >1" thick
> *medium-sized*: cap 2–7" broad; top of stalk >⅜" thick
> *small*: cap ½–2" broad; top of stalk <½" thick
> *tiny*: cap <¾" broad; stalk <⅛" thick.

Don't attach undue importance to the above numbers; they are merely guidelines. For instance, the candy cap is described as "small" because its cap, although sometimes larger, is typically 1–2" broad.

King bolete "shrumps": when competition is fierce, prized mushrooms must be gathered while still under the duff and visible only as mushrumps (low mounds or humps).

Where. Describes the habitat, frequency, and *known* geographical range of the mushroom in western North America. The season is also mentioned if it differs significantly from the peak mushroom season (see p. *xii*).

Most geographical terms, such as "Sierra Nevada," are self-evident. Some possibly ambiguous ones are:

Pacific Northwest: extreme northern California, Oregon, Washington, Idaho, British Columbia, and Alaska.

West Coast: from the Pacific Ocean up to and including the western slopes of the Sierra Nevada and Cascades

Great Basin: from the eastern slopes of the Sierra Nevada and Cascades east to the Rockies

Southwest: southern Utah, southern Colorado, New Mexico, and Arizona

widespread: found in most or all of western North America where its preferred habitat occurs (has nothing to do with frequency; a widespread mushroom is not necessarily common; *common* means frequently encountered).

Bear in mind that the exact distribution of most mushrooms has not been determined. You are encouraged to report findings outside the known range!

Edibility. The edible or poisonous properties of the mushroom are briefly described. Always remember that an idiosyncratic reaction (usually gastrointestinal upset) can occur after eating *any* edible mushroom. Adverse reactions to edible mushrooms are specifically mentioned only when they occur with higher-than-average frequency. *Not recommended* means it is not advisable to eat the mushroom; for instance, it may be unpalatable (too tough or bitter) or possibly poisonous (insufficient data or unpredictable reactions).

Note. A catch-all for miscellaneous comments, comparisons with other mushrooms, or reiterations of key features. At the end there is a page reference to *Mushrooms Demystified*, 2d ed. (1986), abbreviated MD. Some mushrooms are accompanied by captioned photographs, stories, and recipes on the facing page.

Spore Color and Spore Prints

To identify many gilled mushrooms with certainty it is necessary to know the color of their spores. Since spores are microscopic, their color must be determined in mass, preferably by making a *spore print.* Choose the cap of a mushroom that is producing spores (i.e., neither too young nor too old), place it gills-down on a piece of white paper, and cover it with a glass or bowl. Enough spores should be discharged onto the paper (in a gill-like pattern) that their color can be determined. Spore prints can also be obtained from boletes, teeth fungi, and several other groups.

The major advantage of spore prints is their reliability. Spore color is not subject to the whim of the weather or other environmental influences, as are so many features such as cap color. The major drawback of spore prints is that they take several hours, even overnight, to obtain. You can avoid this delay by starting them in the field: just wrap the cap with an index card in waxed paper and carry it gills-down in your basket. If you can't obtain a spore print, look for spore dust on the stalk, on the ring or veil remnants, on the lowermost caps of a cluster, or even on the gills, and console yourself with the fact that many mushrooms can be safely distinguished *without* ascertaining their spore color.

Collecting Gear

Equipment is minimal: a basket or bucket ("universal pail") to carry the mushrooms; waxed paper or waxed bags (*not* plastic) for separating and protecting the different kinds;

Gary Lincoff, author of *The Audubon Society Field Guide to North American Mushrooms*, always wraps his mushrooms in wax paper so they don't get crushed when he drops his basket or leaves it on top of the car.

a trowel (always dig up unknown mushrooms carefully so as not to overlook features at the base of the stalk such as a volva); a small knife and cloth or brush for trimming and cleaning the ones you know are edible; a cheese sandwich (if you're always hungry like me); a compass and whistle if you're in a remote area; a copy of this book.

Seasonal Occurrence

Mushrooms need moisture to develop; most species prefer temperatures of 40-70°F. In every region there is an optimum period, the "mushroom season," when most of the mushrooms appear. In the far north, high mountains, deserts, and other places with severe climates, the "window of opportunity" is compressed into a few weeks. In milder climates, especially along the coast, it may span several months or more than half the year. The peak mushroom season for each region is given below. Of course, each year is a little different; the season may be "early" or "late" depending on rainfall and temperature patterns.

Alaska: August–mid September
Sierra Nevada, Cascades, Rocky Mountains: late July–early September (after thundershowers)
Lower mountains and foothills of higher ranges: September–November or until the first heavy frosts
Pacific Northwest: September–November
Northern California Coast: October–January
Central California Coast: November–March
Southern California Coast: December–March (but very erratic)
Central Valley of California: October–January
Southwest: late July–September
Deserts: sporadic (whenever it rains).

In many areas, especially those with a snowpack during the winter, a smaller flush of mushrooms appears in the spring. Certain species such as morels are abundant at this time but rare or absent during the main mushroom season.

As the mushroom season unfolds, you will notice a succession or procession of different mushrooms. Just as certain wildflowers bloom before others, the king bolete fruits early in the season, while the hedgehog mushroom appears much later, and the chanterelle can be early *or* late. However, the sequence may vary from region to region and from year to year. Much of the "sport" of mushroom hunting lies in learning the habits of your favorite mushrooms. As in the larger world, *timing is everything*: a few days early, and they will be invisible, still under the mulch; too late, and they will be gone or "occupied by other tenants" (see p. 204).

Tundra delectables from Alaska: delicious milk caps, gypsy mushrooms, orange birch boletes.

Eating Mushrooms

Your interest in mushrooms may have begun with a question humans have asked for centuries: "Is it good to eat?" While your interest needn't end with this question, edibility is certainly a major consideration. For centuries mushrooms have been revered for their exquisite and varied flavors, and they rank among the world's most expensive foods. Black truffles, for instance, cost more than $500 a pound, and one perfect Japanese matsutake can fetch as much as $200!

In fungophobic cultures such as ours, it is often said that mushrooms have no food value. This simply isn't true. Mushrooms are high in protein (on a dry weight basis, they are closer to meat and fish than to vegetables), and they form complete protein when eaten with grains (they contain lysine, which grains tend to be deficient in). Mushrooms are also good sources of vitamins and trace minerals, and are low in calories (fats and digestible carbohydrates). Mushrooms, in fact, are frequently used as main dishes in China and Japan; they are far more than edible afterthoughts.

As you try different kinds you will find that each has a particular fragrance, flavor, and texture. Some, like the chanterelle, are delicate and temperamental; they demand special treatment. Others are robust and good in just about anything. Of course, people's tastes differ, and so does the flavor of a mushroom from one patch to another or from one season to the next. Mushroom hunters will argue endlessly about the merits of one kind versus another, just as wine connoisseurs debate the attributes of pinot noir and cabernet.

I prefer simple, elegant dishes that highlight a mushroom's flavor and texture rather than drowning it in butter or garlic. Space permits the inclusion of only a few recipes in this book, but there are several wild mushroom cookbooks available. Here are some additional guidelines.

1. People can have adverse idiosyncratic reactions to edible mushrooms just as they can to scallops or peanut butter. Since each kind of mushroom is a different food, you can be allergic to one and not to another. To minimize the chances of an adverse reaction, cook each kind well and eat a modest amount the first few times.

2. With a few exceptions, mushrooms should not be eaten raw. They are safer, more digestible, and more nutritious when cooked. If you saute them, it is much healthier to use olive oil than butter or cream.

3. Always identify what you eat. Just because a book tells you a mushroom is edible doesn't mean you should eat it. Except for a few "goof-proof" exceptions like the lion's mane or corn smut, it is wise to collect an edible mushroom *several* times before eating it. Once you are on a first-name basis with it, as thoroughly familiar with it as you would be with an old friend, *then* cook it up.

4. *When in doubt, throw it out!* If you eat a harmless mushroom while harboring lingering or unvoiced doubts about its edibility, you may suffer an anxiety attack that mimics the poisoning symptoms you fear.

5. Always remember where you find an edible mushroom. Since the fungal mycelium that produces the mushroom is often perennial, you can return later the same season or the next year for more. A major part of mushroom hunting is discovering and accumulating your own "patches" that produce mushrooms regularly.

6. Pick only fresh, firm specimens. With a few notable exceptions like the shaggy mane, they can be safely refrigerated for several days, but the fresher they are, the better they will taste. Common sense dictates that you don't eat old or rotten mushrooms, or those that seethe with fat, agitated maggots.

7. Don't reject a mushroom just because it disappoints you the first time you eat it. The fault may be yours. The famed matsutake, for instance, loses its complex fragrance and flavor if sauteed like a chanterelle, and the texture of a bolete depends both on how you cook it and the conditions under which it grew. Most renowned mushrooms deserve their reputations.

8. Before refrigerators were invented, people took advantage of mushrooms' fleeting abundance by salting, pickling, or drying them for year-round use. Nowadays, drying and freezing are the most popular methods of preservation. You can dry them in a food dehydrator or home-made dryer (a large box with screens and light bulbs), over a wood stove

or other dry heat source, or, weather permitting, in the sun. Some species, such as the chanterelle, remain leathery after being reconstituted, but many others revive nicely, notably morels (they're hollow), the horn of plenty, candy cap, fairy ring mushroom, and various boletes (which must be sliced first and checked for maggots). Many mushrooms can be frozen after first being sauteed. Raw boletes can be frozen whole (providing you have a freezer big enough to accommodate their bulk), but should be sliced and sauteed *before* they thaw out completely.

Medicinal Mushrooms

Mushrooms are used medicinally by several cultures, but nowhere more than in China. Fungi, especially polypores and other wood-rotting species, occupy a central role in the traditional Chinese pharmacopeia. The *ling chih (Ganoderma lucidum*, pp. 192–193) is believed to promote well-being and

Chaga (*Inonotus obliquuus*) grows on birches throughout the northern hemisphere, forming large, very hard, deeply cracked black cankers that are yellow brown within. Tea prepared from chaga is widely used by Siberians and Russians (like Olga, shown here) to enhance the immune system and combat or prevent cancer. To make chaga tea: cut into pieces with an ax and soak for 6-8 hours, then powder the soaked chunks by running them through a meat grinder. Bring 5 cups of water to a boil and stir in 1 cup of ground chaga. Steep for an hour or two, then filter and drink, reheating it if desired (excess can be kept in a thermos). The tea is richly colored with a very mild flavor.

longevity, and is a recurring motif in Chinese art. One of several species thought to stimulate the immune system, *Grifola umbellata*, has shown powerful cancer-inhibiting effects in double-blind experiments conducted in China. Both the tree ear and the shiitake lower blood cholesterol, and the latter also has anti-viral properties.

Since penicillin and several other "wonder drugs" are derived from fungi, one would expect mushrooms to be investigated vigorously by western medicine. Instead, they have been neglected, a tribute to the "worthless at best" aspect of fungophobia, a cultural inability to perceive mushrooms as potentially powerful. Also, there is a strong prejudice against "panaceas" as opposed to drugs with specific effects (see Andrew Weil's comments on p. 193). For more information, consult *Fungi Pharmacopoeia (Sinica)* by Liu Bo and Bau Yun-sun (Oakland: Kinoko Company, 1980).

Consciousness-altering Mushrooms

Mushrooms are among the oldest known entheogenic (consciousness-altering or hallucinogenic) agents. The most famous are the fly amanita (*Amanita muscaria*) and those that contain psilocybin. The fly amanita was used widely in northern Eurasia, from Siberia to Scandinavia, and by some Native American peoples. It may also have been the mystical "Soma" praised so loftily in the sacred Vedic texts of the ancient Aryans. Unfortunately, this mushroom and its relative the panther amanita contain varying and unpredictable amounts of several different toxins, and individual reactions to them differ enormously. As vomiting, profuse sweating, delusions, and even convulsions can result, their use is not recommended.

The psilocybin- and psilocin-containing mushrooms, most of which belong to the genera *Psilocybe* and *Panaeolus*, produce effects similar to those of LSD but of slightly less duration. These mushrooms are revered in southern Mexico, where there is a long shamanistic tradition of ingesting them for religious, healing, and divination purposes. In spite of the widespread hysteria surrounding drug use (as opposed to drug abuse), psilocybin-containing mushrooms are still popular recreational drugs. Several species, however, are easily confused with deadly poisonous mushrooms, and it is illegal to possess them.

To learn more, consult *The Wondrous Mushroom: Mycolatry in Mesoamerica* by R. Gordon Wasson (New York: McGraw-Hill, 1980); *Soma: The Divine Mushroom of Immortality* by R. Gordon Wasson (New York: Harcourt Brace Jovanovich, 1968); and *Psilocybin Mushrooms of the World* by Paul Stamets (Berkeley: Ten Speed Press, 1996).

Poisonous Mushrooms

The dangers of eating mushrooms are greatly overstated, even by mushroom experts (a tribute, once again, to our cultural antipathy toward mushrooms). Once you learn how to avoid the few truly dangerous kinds, the greatest risk you face is gastrointestinal distress of a greater or lesser degree, and our bodies are built to handle such distress. Yet we are inundated with dire warnings about the dangers of "toadstool testing" that are vastly disproportional to the real damage that mushrooms inflict (they cause very few if any deaths annually in North America).

The most dangerous mushroom toxins are the amanitins found in the deadly amanitas and several LBMs ("little brown mushrooms"). Learn how to recognize these mushrooms before you learn the edible ones. The symptoms are typically delayed for 6–24 hours after ingestion, so if you have reason to believe that someone has eaten a deadly mushroom, *seek medical attention immediately!* Don't wait for the symptoms to appear; by the time they do the toxins will have been absorbed.

Should you be poisoned by a mushroom, try to remain calm, identify the agent responsible, and seek medical attention if you need it. Idiosyncratic reactions to edible mushrooms are generally not serious enough to warrant a trip to the hospital, but if there is any doubt, err on the side of safety. Also remember that "mildly poisonous" mushrooms can have more serious effects on small children.

For a more detailed listing of poisonous mushrooms and their effects (including some kinds not depicted in this book), see MD 892–896 and *Toxic and Hallucinogenic Mushroom Poisoning* by Gary Lincoff and D.H. Mitchel (New York: Van Nostrand Reinhold, 1977).

Growing Mushrooms at Home

As knowledge of how to grow fungi has "mushroomed," so has interest in "taming" wild species. As a rule, any mushroom that grows on wood or humus is a good candidate for cultivation, although some are more cooperative than others. Among the easiest to grow are the oyster mushroom, king stropharia, shiitake (a native of Japan), species of *Agaricus* and *Hericium*, and the sulfur shelf. Even morels can be grown at home with a little effort — and a lot of luck.

Mycorrhizal mushrooms, on the other hand, are extremely difficult to raise without their preferred tree hosts, but delicious species like the chanterelle, king bolete, and matsutake may someday be grown by landowners as a cash crop while they wait for trees to reach harvestable size. For more infor-

Home-grown mushrooms (clockwise from top left): shiitake, two strains of oyster mushrooms, ling chih, lion's mane, enokitake.

mation on growing mushrooms, see *Growing Gourmet & Medicinal Mushrooms* by Paul Stamets (Berkeley: Ten Speed Press, 1993).

Dyeing with Mushrooms

Mushrooms yield a lovely range of color-fast natural dyes. Some of the best dye mushrooms are the western jack o' lantern (purple), *Dermocybe* species (red, rose, purple, orange), *Hydnellum* and *Sarcodon* species (blue and green), and the dyer's polypore (yellow, orange). The procedure is simple: simmer wool (pre-mordanted if desired) in a dye bath with the mushrooms. Mushrooms can also be used to dye hair (see p. 145). Miriam Rice, who pioneered the use of mushroom dyes, has written an excellent book on the subject, *Mushrooms for Color* (Eureka: Mad River Press, 1980).

Mushroom-dyed wool.

Yana, Anya, Vova, and Vadik: young Russian mushroom hunters (see p. *xxi*) with panther amanitas.

There isn't room in this book to discuss all the uses of mushrooms. People play games with them (p. 217), draw on them (p. 196), dance with them, or wear them (as shown here). Mushrooms have also been used to start fires, smoke out bees, poison enemies, and plug leaky roofs. And of course, they have ecological uses: mycorrhizal mushrooms are crucial to the success of reforestation projects, and certain species have been shown to break down toxic wastes. **Right**: Al Catalini models a traditional Romanian cap made from pounded conks (polypores). **Below**: Foraging fashions: more mushroom caps.

Brandi Lyn Scates with a shaggy parasol (*Lepiota rachodes*). "What could be more thrilling and fulfilling than finding a nice big mushroom in the woods?" asks lepiota-lover Grainger Hunt (below).

Mushrooms and Kids

Kids make terrific mushroom hunters: naturally curious, highly energetic, low to the ground, and remarkably discerning — sometimes more so than adults. Mystery writer Lia Matera relates this story of her seven-year-old son:

> Brendan *loves* your book [*Mushrooms Demystified*]; he spends hours looking through it, memorizing all the mushrooms, and he draws pictures of them which he assembles into little mushroom books of his own. One day we were in a bakery with paintings of different mushrooms on the walls. I pointed to a beautiful white, ringed mushroom and said, "Brendan, that's one you *don't* want to eat; it's called the destroying angel." Brendan studied it carefully and said, "No it isn't, it's a white matsutake." Well, I'd never seen a white matsutake, but when we checked the book, he was right!

Except for toddlers in the grazing stage, children are unlikely to eat something unless they know it's edible, especially if they are also taught what is poisonous. Yet most North Americans are terrified by the prospect of their children taking an interest in or, even worse, *handling* wild mushrooms. (Fact: relatively few mushrooms are poisonous, and they can only cause harm if ingested.) Typically, the only kind of contact that is encouraged is the kicking and stomping of "toadstools" that sprout on lawns. Grown males can often be seen leading the charge. Even adults who have taken mushroom classes are frequently reluctant to share their newfound knowledge with their offspring — a case of *projection*, not protection, if you ask me. (In a society that delegitimizes adult feelings on the one hand and encourages, even exalts, parental concern for innocents on the other, it is not surprising that adults commonly project their own fears onto their children, in this case, a lingering residue of distrust from their fungophobic upbringing.) Contrast this climate of overbearing concern with Valentina Wasson's description of her childhood in Russia (from *Mushrooms, Russia, and History*; New York: Pantheon 1957):

> What a delight it was to ramble through the clean, fragrant woods, filling our baskets. [When I was almost eight and my sister was nearly seven] . . we were already proficient mushroom gatherers, and we must have begun our apprenticeship long before. Our mother was even more solicitous about her brood than most mothers, yet it never occurred to her to poison our young minds with warnings about "toadstools." All Russians know the mushrooms, not by dint of study as the mycologists do, but as part of our ancient heritage, imbibed with our mother's milk . . . Tanya and I, and all our little playmates, made ourselves useful, when in the country, by gathering various kinds of mushrooms and bringing them home in childish rivalry and glee to the kitchen. When we were naughty, our mother would punish us by forbidding us to go mushrooming.

Many immigrants to North America raised their children as mushroom hunters. Italian-American John Feci (see p. 156) remembers when he was ten:

> My mother was the expert in the family. She knew I could cover a lot of ground, so she would encourage me to go mushroom hunting. I'd bring home a bucket of boletes and get a lot of praise. I loved finding them, but I also did it for the recognition, to get in her good graces because she was a tough mom.

Biologist Grainger Hunt, a lepiota lover and father of three, amplifies the theme of children making themselves useful:

> How easily we forget that we evolved as hunters and gatherers, and that our bodies are marvelous machines designed for those purposes. Kids were once *indispensable* to the survival of families. When they brought home their first bird or mushroom, it was a *major* event, their mothers got really excited because it meant the kid had become a *producer* — not of trivial things, but of *food*. Now so many kids lack a vital role, they question their relevancy because they have no function in the family. *I think we should send them out mushroom hunting.* Can you imagine the *excitement* kids feel when told to find five pounds of luscious lepiotas for dinner? Can you imagine the sense of *meaning*, and *purpose*, that gives them? They get to be on a *mission*, to learn about the world in a challenging and purposeful way that fits perfectly with the way they're designed. I mean, what could be more thrilling and fulfilling than finding a nice big mushroom in the woods?

What, indeed? North America will remain a continent of fungophobes until its children pursue mushrooms as freely and fearlessly as do their Russian counterparts.

"Before the concert." Mushroom lovers are opportunivores, always ready to grab a meal when one presents itself. Five minutes before a chamber music concert, this trombonist was spotted gleefully gathering chanterelles outside the concert hall.

Boletes or Chanterelles?

From a picking perspective, since that's what I do, it's obviously chanterelles, because I've picked them in more areas than I'll ever pick boletes. When boletes come to mind it's always *Boletus edulis*. As far as being a boletivore, I'm one of the most avid. But I would have to say chanterelles because you can pick them in more areas, and because of their beauty. But then again, boletes have a *tremendous* beauty about them. There's never been a bolete I didn't like. But then again, there's never been a chanterelle I despised either. So that will have to do it: chanterelles when they're available and boletes when they're available.

—*Mohammed Ismail*

Chanterelles. It's the color and the fruitiness. I like boletes very much, but my mouth waters for chanterelles.

—*Greg Wright*

Chanterelles.   —*Harold Furuta*

xxii

Above: A batch of pig's ears (*Gomphus clavatus*). **Right**: Andrew Weil with a basket of hefty chanterelles; note how pale the gills are in this California variety. **Below**: This smaller, brightly colored version of the chanterelle is common in the Southwest. Note how the gills are forked and veined.

1 Chanterelles

In this group the cap is usually vase-shaped, trumpet-shaped, or wavy at maturity, and the spore-bearing surface that lines the underside is completely smooth, shallowly wrinkled, or furnished with primitive gills. Unlike the thin blades of the gilled mushrooms, chanterelle gills, when present, tend to be fairly thick and shallow, blunt, and often connected by cross-veins (see above photo). The stalk is fibrous (it doesn't snap open cleanly like chalk), the spores are white to yellowish or ochre, and there is no veil.

Chanterelles rank among the most popular edible wild mushrooms. If your "chanterelle" is not among those described here, check the flat-topped club coral (p. 212) and lobster mushroom (p. 248), and see MD 658–668.

Chanterelle (*Cantharellus cibarius*)

Other Names: Yellow Chanterelle, Girolle, Pfifferling.

Key Features:
1. Cap bright orange to yellow-orange, bald, usually concave or wavy when mature.
2. Gills well spaced, shallow, blunt-edged, and fairly thick, often with connecting veins in between.
3. Gills same color as cap or paler, running down the stalk.
4. Stalk colored like cap or slightly paler, solid (*not* hollow).
5. Flesh white or with a tinge of yellow.
6. Growing on the ground.

Other Features: Medium-sized to large; cap broadly domed to nearly flat when young; odor usually fruity (like pumpkin or apricot) but sometimes mild; veil, ring, and volva absent; spores yellowish.

Where: On ground under conifers and oaks, often in groups but *not* in fused clusters; abundant from southeast Alaska to California and the Southwest.

Edibility: Very popular because of its cheerful color, delicate flavor, and fruity fragance. It is usually free of maggots.

Note: The chanterelle's size, shape, and even its color vary from region to region. Those in California are the largest in the world, while the Rocky Mountain version is much smaller and more brightly colored (see photos on p. 1). The poisonous western jack o' lantern mushroom has thinner, more crowded gills and does *not* have white flesh; the false chanterelle has thinner, oranger gills and often a browner cap. See MD 662–664 and plates 175, 177–178.

Cream of chanterelle soup is a traditional favorite. You can use a basic cream of mushroom soup recipe and vary it according to taste. It is best to dry-saute the chanterelles *before* putting them in the soup.

On Cooking Chanterelles

Chanterelles are often dirty, and when washed they soak up water like a sponge. All too often the result is disappointing, if not downright disastrous (see p. 70). But there is a simple way to cook chanterelles so they don't become too spongy or slimy. It's called dry-sauteeing (see p. 55 for details), and it concentrates their flavor while allowing you to wash them as vigorously (or as savagely) as you want. Once they've been dry-sauteed, you can incorporate them into any dish, including that traditional favorite, cream of chanterelle soup.

Clean, pristine chanterelles like those don't require washing.

"If there's one thing I know," bragged El Bo
"It's where all the best chanterelles grow."
But when he went the next fall
What he found did appall
Condominiums sprouting all in a row!
—Charles Sutton

Boletes or Chanterelles?

Boletes are the round mother-earth mushrooms of the forest floor. They're rich, they're nutty, they're buttery, and their flavor is the flavor of the forest. Chanterelles are more like the queen seductress: fruity, peppery, richer, more difficult to work with from a cooking standpoint, and complex and very singular. I don't have a preference. They're so different. It's like comparing pinot noir and cabernet — whichever one you happen to like is better!
—Jack Czarnecki

My Most Memorable Mushroom Hunt

It's tough to single out just one, because the mushroom that gives me the greatest pleasure is unquestionably *Agaricus augustus*. But I've never filled a basket with it. It's always been a *lot* of looking and one day I get a few.

I guess the day that surpassed all others was when there were so many white chanterelles it began to irritate us to find them. We'd started out very methodically, looking for just the *right* spots and the *right* trees, and pretty soon the four of us filled our baskets. Then, as we were walking back to the car, completely satiated with the whole experience, we found that all these funny feeling spots we kept stepping on were just *more* of them. We couldn't take a step without finding another one. It was memorable, but irritating!
—Lester Hardy

White Chanterelle
(*Cantharellus subalbidus*)

Key Features:
1. Entire mushroom dull white (but bruised areas usually yellowish or orangish).
2. Cap usually concave or wavy when mature, bald.
3. Gills running down the stalk, well spaced, shallow, blunt-edged, and fairly thick, often with connecting veins in between.
4. Stalk solid (*not* hollow), *not* snapping open cleanly like a piece of chalk.
5. Veil, ring, and volva absent.
6. Growing on ground in woods, *not* in fused clusters.

Other Features: Medium-sized to large; flesh thick and white; spores white.

Where: On ground in woods, often hiding under the humus; common from the Pacific Northwest to central Californa. It favors pine and other conifers but also grows under tanoak and manzanita.

Edibility: Popular; some prefer it to the chanterelle.

Note: Like the chanterelle, which it resembles in all respects except color, this meaty mushroom is not readily attacked by maggots. There are many large whitish mushrooms with gills that extend down the stem, but their gills are thin-edged like the blade of a knife, *not* blunt-edged or foldlike. See MD 662 for more information.

Yellow Foot
(Cantharellus tubaeformis)

Other Names: Winter Chanterelle, Funnel Chanterelle, *Cantharellus infundibuliformis*.

Key Features:
1. Cap small (mostly <2" broad), usually with a central depression.
2. Cap dark brown to tan or dull orange.
3. Gills few and widely spaced, running down the stalk.
4. Gills thick, shallow, blunt, often connected by veins.
5. Stalk slender (± ¼"), orange to yellow when fresh.
6. Stalk hollow at maturity (and often when young).

Other Features: Cap *not* sticky or slimy; gills yellowish to orangish in some forms, grayish to brownish or violet-tinged in others; stalk often fading; veil, ring, and volva absent; spores white to yellowish.

Where: On ground, moss, or rotten wood in wet conifer forests or bogs, often in large numbers; widespread in northern latitudes, common from Alaska to San Francisco.

Edibility: Edible. Not as meaty as other chanterelles, but with a good flavor, especially when dry-sauteed.

Note: This is the most common chanterelle of bogs and cool mossy forests. In California it fruits in the winter, often accompanied by hedgehog mushrooms. Several waxy caps (*Hygrocybe*) are similar but have thin-edged gills. See MD 665 and plate 180.

Pig's Ears (*Gomphus clavatus*)

Other Names: *Cantharellus clavatus.*

Key Features:
1. Mature cap yellow-brown to tan or olive-tinged, the edges usually turned up and wavy.
2. Surface of cap *without* prominent scales.
3. Underside purplish or at least partly purple when fresh, with wrinkles or ridges that extend down the stalk nearly to the base.
4. Stalk solid (*not* hollow).
5. Flesh white.

Other Features: Medium-sized; often more or less club-shaped when young (before the cap broadens); cap sometimes purplish when young; one side of cap usually higher than the other; spores pale tan or ochre.

Where: On ground or rotten wood in old mossy conifer forests or in mixed woods, usually clustered and often forming large rings or arcs; widespread and common in northern latitudes, southward in mountains, and along the coast to San Francisco.

Edibility: Very good when young, but often riddled with maggots.

Note: This mushroom dyes wool a lovely lavender when used with an iron mordant. The purple wrinkled underside is its most distinctive feature. Young clublike specimens are reminiscent of club corals (*Clavariadelphus*), while older individuals are shaped more like chanterelles. The name "pig's ears" is also applied to an unrelated group of springtime cup fungi, *Discina*. See MD 661 and plate 176.

Scaly Chanterelle (*Gomphus floccosus*)

Other Names: Woolly Chanterelle, Vase Chanterelle, *Cantharellus floccosus*.

Key Features:
1. Mushroom more or less vase-shaped, the center of the cap sunken or even hollow.
2. Cap red to orange.
3. Surface of cap with prominent scales (unless very young or battered by rain).
4. Underside of cap whitish to yellowish or pale ochre, with veins or wrinkles running down the stalk.

Other Features: Medium-sized to fairly large; veins on underside sometimes forming wide, shallow pits; veil, ring, and volva absent; spores ochre.

Where: On ground near or under conifers, often in groups or clusters; common in the Pacific Northwest and coastal northern California, and in most mountain ranges.

Edibility: Not recommended; many are made ill by it.

Note: The scaly chanterelles are a prominent feature of our northern and mountain conifer forests. *G. floccosus* is the most colorful of the lot; the others range from pale orange (*G. bonari*, facing page) to tan or brownish (*G. kauffmanii*, below). See MD 661–662 and plate 174.

Gomphus bonari closely resembles *G. floccosus*, but has a slightly duller cap (pale orange to orange-buff), it is common under mountain conifers.

My Most Memorable Mushroom Hunt

I will always remember spending two solid days cooped up in a car with Gregg Ferguson driving to God Knows Where in New Mexico to find nothing — and coming back to tell about it. I'll never forget it, the biggest bust I've ever had. We spent a week. I don't think I brought anything home. We ate what we found. We did have one memorable meal: rattlesnake and chanterelle stew. It was tasteless. The Nutter Butters on the rim of the Grand Canyon were poor compensation.

—*Brad Beebe*

I've lived in the mountains of New Mexico for many years, on next to nothing. All mushroom hunts are memorable to me, because I live with no "entertainment." Distractions are not necessary. I live in The Forest. I remember, thus I need no camera or computer. I have nothing against those things; but they have nothing *for* me. Some years I can smell the mushrooms in the night through the open doorway at the foot of my bed. I easily fall asleep recalling the look of the forest floor as I mentally retrace my steps of the day. A memorable mushroom trek may turn up no mushrooms, but give me other beauties, filling me with a sense of adoration and wonder. I set out dumb and return dumbfounded. I try to be aware, a learning person, and I use references. But first and foremost I am *in respectful love* with The Forest.

—*Muyu Szpakowski*

Black Trumpet
(*Craterellus cornucopioides*)

Other Names: Black Chanterelle, Trumpet of Death, Horn of Plenty.

Key Features:
1. Mushroom trumpetlike when young, petunia-shaped when older; cap with a central hollow extending nearly to base of stalk.
2. Cap black when moist, brown in dry weather.
3. Underside of cap gray, smooth to very slightly wrinkled but *without* gills or prominent veins.
4. Flesh tough, thin, *not* gelatinous.
5. Stalk gray, brown, or black (*not* yellow or orange).

Other Features: Medium-sized (mostly 2–5″ high); edge of cap turned down when young, then unfurling and often wavy or frilly in age; flesh dark; stalk hollow; spores whitish to yellowish.

Where: On ground in forests, usually in groups or clusters; abundant from central California to southern Oregon, less frequent northward. It fruits late in the season, often next to fallen branches or sticks.

Edibility: Delicious — the most flavorful of the chanterelles. The tough texture requires prolonged, gentle cooking.

Note: Despite its abundance, this somber-looking mushroom is very difficult to see in the dark forests where it thrives. Visualizing black (or brown) petunias may help you to spot them. A nearly identical edible species, *C. fallax*, has ochre- or salmon-tinged spores. Several cup fungi are black but more cuplike. See MD 666–668 and plate 182.

Sauteed *Craterellus* is good in just about anything, but its black color stands out especially well in omelets, over pizza, or on open-faced sandwiches garnished with slices of tomato, bell pepper, and cheese. It is also excellent alone.

My Most Memorable Mushroom Hunt

It was a bittersweet experience. Bittersweet in that I found, in a place dear to my heart, a spot where there were these mushrooms and *they were all over the place*! But they were beyond the pale. They didn't go *in* the pail, either. What were they? *Boletus edulis*, and *no*, I won't tell you where.

But as long as we're in this vale of tears, I'll tell you something bitter. Not just bittersweet, but bitter: *I haven't actually found those little black trumpets*. Ever! It's shocking, I know, but I'm admitting it now, baring my breast.

—*Scott Anderson*

It was when I found my first *Craterellus*. What a revelation — suddenly they were *everywhere* and I had been looking for them for months. It was just up the hill from my house, so I had gone over the same terrain any number of times. What made the difference? Being told how hard they were to see, how they looked like holes.

—*Anne Shelly*

Blue Chanterelle
(*Polyozellus multiplex*)

Other Names: Black Chanterelle, Clustered Blue Chanterelle.

Key Features:
1. Growing in clumps with several to many lopsided, fan-shaped or spoon-shaped caps whose stalks are usually fused at the base.
2. Cap and stalk deep blue, black, or violet-black.
3. Underside of each cap with wrinkles, veins, or very shallow, blunt gills that run down the stalk, colored like the cap or slightly paler.
4. Growing on ground in northern or mountain forests.

Other Features: Flesh gray to blue-black or black, rather tough; caps *not* hollow at the center; spores white.

Where: In clumps on ground under spruce, fir, hemlock, or sometimes aspen; Alaska to extreme northern California and south through the Rocky Mountains to the Southwest; common in some areas.

Edibility: Edible. It cooks up black but delicious.

Note: This is one of several wild mushrooms marketed under the name "black chanterelle" (the others are *Craterellus* species). The Rocky Mountain version is usually deep blue or violet-tinged, while those from coastal Alaska are more likely to be jet-black with a dark gray underside. See MD 665–668 for more information.

In the Rocky Mountains and Southwest, *Polyozellus multiplex* is often more bluish or violet than in Alaska.

My First Mushroom Hunt

Today was my first mushroom hunt and guess what we found: chanterelles beneath a bay tree! It was just *loaded* — there must have been *pounds* of them! I'd seen them in the store, but never in the woods. My mother used to go mushroom hunting when we were kids. She only picked certain ones and nothing else. Actually, my first mushroom hunt was for psychedelic mushrooms on Maui.

I didn't expect to walk this hard. I mean, we were walking at a *fast* pace. The leader knew we had to get to a certain place, but *I* didn't know that, so I was thinking, "How are we gonna find mushrooms when we're rushing like this?"

But it was the most wonderful experience. I really loved it. And then tasting them — one was like shrimp, another tasted like chicken, and the chanterelles, which I'd never eaten — unbelievable! I'd never eaten the oyster ones either.

—*Rosina Paris*

In *Craterellus cinereus*, another delicious "black chanterelle," the underside of the cap has primitive gills rather than being smooth or slightly wrinkled. It grows in California under oak and tanoak.

The dark red juice of the bleeding milk cap (see facing page).

The short-stemmed russula (p. 27) is one of our commonest woodland mushrooms. Note how the stalk breaks cleanly like a piece of chalk.

2 Milk Caps and Russulas

The outstanding feature of these gilled mushrooms is their texture: the gills and flesh are brittle, and the stalk snaps open cleanly like a piece of chalk (see above photo) unless riddled with maggots. The spores are white to yellow or ochre, and there is no veil. Fresh milk caps often exude a milk or juice (top photo) when cut; russulas do not.

The taste of the flesh (whether mild or peppery) is a helpful feature in this group. Any milk cap or russula can be safely tasted by chewing on a piece of the cap and then spitting it out. Several milk caps and russulas are excellent edibles; a few are mildly poisonous. They grow in the woods or near trees, usually on the ground, and are very common. Only a small portion of the total number is offered here. See MD 63–103 for a more detailed treatment.

Bleeding Milk Cap
(*Lactarius rubrilacteus*)

Other Names: Red-juice Milk Cap, *Lactarius sanguifluus*.

Key Features:
1. Cap bald, reddish-brown to orangish to tan, often with darker and lighter concentric zones, and often with greenish stains (especially when old).
2. Gills reddish, purplish-red, or tan (sometimes with greenish stains), broadly attached to stalk or running down it.
3. Stalk colored more or less like cap.
4. Entire mushroom brittle, the stalk snapping open cleanly like a piece of chalk (unless maggoty).
5. Flesh exuding a dark red juice when cut, or at least showing traces of dark red color.
6. Bruised areas staining green within several hours.
7. Veil, ring, and volva absent.
8. Associated with conifers.

Other Features: Medium-sized; cap with a sunken center when mature; taste *not* peppery; stalk at least ⅜" thick, sometimes with a miniature green mushroom at base; spores buff or pale yellowish.

Where: On ground near or under conifers (mainly Douglas-fir); widespread, but especially common on West Coast.

Edibility: Fairly good, at least better than *L. deliciosus*.

Note: The dark red juice (see photo on facing page) and green staining of all parts serve to identify this mushroom. It is often harmlessly confused with the delicious milk cap (*L. deliciosus*), but the gills and juice are redder. See MD 68–69 for more information.

15

Delicious Milk Cap
(*Lactarius deliciosus*)

Other Names: Saffron Milk Cap, Orange-juice Milk Cap.

Key Features:
1. Cap bald, orange to orange-brown or sometimes grayish, but often stained green (especially when old).
2. Gills bright to dull orange (sometimes with greenish stains), broadly attached to stalk or running down it, *not* promiscuously forked or veined.
3. Stalk orange or orangish, or sometimes stained green.
4. Entire mushroom brittle, the stalk snapping open cleanly like a piece of chalk (unless maggoty).
5. Flesh exuding a bright orange juice when cut, or at least showing traces of bright orange color.
6. Bruised areas staining green within several hours.
7. Veil, ring, and volva absent.
8. Associated with conifers.

Other Features: Medium-sized to fairly large; cap with a sunken center when mature, often zoned concentrically; taste mild or bitter; juice in some variants turning dark red after exposure before eventually staining green; stalk >⅜" thick; spores buff or yellowish.

Where: On ground near or under conifers (mainly pine and spruce), often barely poking above the humus or visible only as a "shrump"; widespread and abundant.

Edibility: Edible but not necessarily delicious. The most common variant is slightly bitter.

Note: When the carrot-orange juice is not evident, the telltale green stains usually are. The variant in the photographs is also called *L. deterrimus*, to distinguish it from the *L. deliciosus* of Europe, which does not stain green as dramatically. See MD 68 and plate 2.

Lactarius deliciosus is abundant under spruce as well as pine.

Salted Milk Caps

Every summer the Russian and Siberian taiga teems with "crock hunters" gathering mushrooms for salting. Milk caps (*Lactarius*) treated in the following way are especially favored because of their crunchiness.

Clean the milk caps, check for maggots, and trim off the stems or at least the bases so that the caps can lie flat. Layer the caps in a crock or wide jar with the gills facing up, and sprinkle each layer with rock salt (1½-2 T. salt per pound of mushrooms). You can also sprinkle every other layer with an herb mixture (black currant leaves are traditional, but a little dill, black pepper, and caraway seeds can be used) or with chopped, cooked onion or garlic (2 cloves per pound). Place a round wooden board or plate on top of the mushrooms and weight it with a stone or jar of water (tightly closed), then refrigerate or store in a cool place. The mushrooms will gradually release their own liquid; excess juice can be poured off and new layers of mushrooms added as they are gathered. Mild milk caps such as *L. deliciosus* and *L. rubrilacteus* can be eaten after only three days. The indigestible peppery ones such as *L. torminosus* (p. 19) and *L. resimus* (see MD 74) must stand for 45 days. If the finished product is too salty for your taste, you can wash the salted mushrooms before eating them.

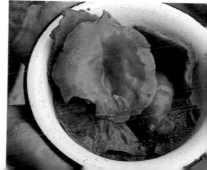

Salted *ryshik (Lactarius deliciosus)* are a favorite Russian snack; they are often served with vodka.

Indigo Milk Cap (*Lactarius indigo*)

Key Features:
1. Entire mushroom blue to blue-gray (sometimes with greenish stains when old).
2. Entire mushroom brittle, the stalk snapping open cleanly like a piece of chalk (unless maggoty).
3. Flesh exuding a bright blue or dark blue juice when cut, or at least showing traces of blue color.
4. Veil, ring, and volva absent.

Other Features: Medium-sized; cap bald, often grayer or paler than gills and often with concentric zones, the center sunken when mature; odor *not* aniselike; bruised areas slowly staining green (within several hours); stalk usually at least ⅜" thick; spores yellowish or creamy.

Where: On ground under pine and oak in Arizona and possibly New Mexico, but not nearly as common there as in the southeastern United States.

Edibility: Tasty as well as colorful. In Mexico it is sometimes sold in farmer's markets.

Note: The blue color of this mushroom makes it virtually unmistakable. Another edible milk cap, *L. chelidonium*, often shows a little blue in the cap but has brown to yellow-brown juice (when present); it is fairly common in the Southwest. See MD 69 and plate 4.

Bearded Milk Cap
(*Lactarius torminosus* & *L. pubescens*)

Other Names: Woolly Milk Cap, Pink-fringed Milk Cap.

Key Features:
1. Cap whitish to pale pink or pinkish-orange.
2. Cap with a sunken center, the edge bearded with soft, woolly hairs and rolled under when young.
3. Gills whitish to cream or tinged pink, close together.
4. Stalk short, firm, colored like cap or paler.
5. Entire mushroom brittle, the stalk snapping open cleanly like a piece of chalk.
6. Taste intensely peppery (chew on a small piece of cap, then spit it out).
7. Spores white or creamy.
8. Associated with birch.

Other Features: Medium-sized; gills when very fresh exuding a milk when cut, attached to stalk; milk if present white, unchanging or staining yellowish when exposed; stalk hardly spotted if at all; ring and volva absent.

Where: Widespread on ground near planted birches, and in northern forests wherever birch occurs; common.

Edibility: Special methods, such as salting or pickling, are required to remove the acrid taste and avoid stomach upsets. For advice, consult the Lithuanian or Russian nearest you.

Note: Whether or not the milk is evident, these two species (which differ microscopically) can be recognized by their color, bearded cap with a sunken center, and burning taste (which can blister the tongue if sampled too liberally). Other bearded milk caps grow elsewhere and/or are differently colored. See MD 73 for more information.

Purple-staining Bearded Milk Cap
(*Lactarius representaneus*)

Other Names: Northern Bearded Milk Cap.

Key Features:
1. Cap yellowish to orange-yellow (or with purple stains), usually with fibers or hairs toward edge, sticky when moist.
2. Edge of young cap rolled under and bearded with woolly yellowish hairs.
3. Gills buff to yellowish (but often with purple stains), exuding a white milk when cut.
4. Bruised areas slowly staining dull purple.
5. Entire mushroom brittle, the stalk snapping open cleanly like a piece of chalk.
6. Veil, ring, and volva absent.
7. Associated with conifers.

Other Features: Medium-sized; cap sometimes with concentric zones, sunken at center; taste usually bitter or peppery; gills attached to stalk; stalk hard, colored like cap or paler, usually spotted or pitted; spores white to yellowish.

Where: Widespread and common on ground under northern and mountain conifers (especially spruce).

Edibility: Not recommended.

Note: The yellowish bearded cap and purple staining of all parts are distinctive. Other purple-stainers, such as *L. uvidus* and *L. pallescens*, are neither bearded nor yellow. Still other milk caps, such as *L. scrobiculatus*, are bearded and yellow but do not stain purple. See MD 73–76 for more information.

Red Hot Milk Cap (*Lactarius rufus*)

Key Features:
1. Cap and stalk brick-red to red-brown, *not* sticky or slimy.
2. Cap bald, *not* zoned concentrically.
3. Gills white or tinged cap color, exuding a milk when cut (unless old, soggy, or dry).
4. Milk white, *not* changing color when exposed.
5. Entire mushroom brittle, the stalk snapping open cleanly like a piece of chalk.
6. Taste intensely peppery (chew on a small piece of cap, then spit it out).
7. Veil, ring, and volva absent.
8. Associated with conifers.

Other Features: Medium-sized; mature cap flat to slightly concave, sometimes with a small knob at center; spores white or pale yellowish.

Where: On ground or in moss near or under conifers (especially pine and spruce), usually in groups or troops. Widespread and common from the forests and bogs of Alaska to northern California, and southward in mountains.

Edibility: Not recommended; it may be slightly poisonous.

Note: The dry, dull reddish cap and stalk, white milk, and red-hot taste are distinctive. Be careful when tasting it, because the hotness is often delayed and becomes more intense with time! See MD 79–80 for more information.

Candy Cap *(Lactarius rubidus)*

Other Names: *Lactarius fragilis* var. *rubidus, L. fragilis.*

Key Features:
1. Cap and stalk burnt orange to orange-brown or sometimes reddish-brown (stalk often slightly paler).
2. Cap small (usually <2½" broad), the surface dull and dry, *never* sticky, slimy, or shiny.
3. Fresh gills exuding a watery white milk when cut.
4. Neither the milk nor the flesh changing color when exposed (*never* staining yellow).
5. Entire mushroom brittle, the stalk snapping open cleanly like a piece of chalk.
6. Veil, ring, and volva absent.
7. Odor sweet (like maple syrup or fenugreek), at least when cooked or dried.
8. Taste *not* peppery (chew on a small piece of cap).
9. Spores white.

Other Features: Mature cap bald, flat to slightly sunken or with a small central knob; gills pale or tinged cap color, broadly attached to stalk or running down it; stalk usually slender, usually hollow at maturity.

Where: On ground or rotten wood in forests, usually in groups; Pacific Northwest to central California. In California it is abundant under oak and pine.

Edibility: Delicious fresh, but sweeter dried; excellent in cookies, breads, waffles, and pancakes.

Note: Fresh candy caps can be recognized by their burnt orange color, dry cap, and watery white milk. Older or drier individuals often lack the milk but have an unmistakable maple syrup aroma. If allowed to dry slowly, a single candy cap will scent an entire house. Several milk caps look similar but are not fragrant. See MD 79–82 for more information.

Candy cap cookies with dried and fresh candy caps.

Candy Cap Cookies

Measure 1 cup dried and chopped candy caps and soak in ¼ to ½ cup lukewarm water for 15 minutes. Then saute in a little butter. If the cooked mushrooms are slightly bitter, add a little water and saute longer to remove the bitterness. Set aside.

Cream together 1 cup soft butter and 1 cup sugar. Beat in 1 egg and, if you like, ½ tsp. vanilla extract. Stir in 2½ cups flour; then add and blend in ½ cup finely chopped raw cashews and the sauteed candy caps.

Form into rolls about 1 to 1½" in diameter. Wrap in wax paper and freeze. When ready to bake, remove rolls from freezer and slice thin, placing on ungreased cookie sheets. Bake at 350° for 8 to 10 minutes or until very lightly browned around the edges. Makes about 6 dozen. Delicious!

—*Rose Marie Milazzo*

The rufous candy cap, *L. rufulus,* is common under oak in southern California. It has yellowish or cream-colored spores and is slightly stouter and redder than the candy cap. It can be used in the same recipes but is not quite as fragrant.

Orange Milk Cap
(*Lactarius luculentus*)

Other Names: *Lactarius aurantiacus.*

Key Features:
1. Cap orange to orange-brown, bald, sticky when moist and slightly shiny when dry.
2. Gills white or tinged orange, exuding a milk when cut.
3. Milk white, *not* changing color when exposed.
4. Stalk orangish (colored more or less like cap).
5. Entire mushroom brittle, the stalk snapping open cleanly like a piece of chalk.
6. Odor *not* distinctive.
7. Taste *not* peppery (chew on a small piece of cap).
8. Veil, ring, and volva absent.
9. Associated with conifers.

Other Features: Small to medium-sized; cap domed when young, flat or concave in age or with a small knob at center; taste mild or bitter; gills broadly attached to stalk or running down it; stalk soon hollow, often long; spores white or tinged yellow.

Where: Common in duff or moss under conifers from Alaska to coastal northern California; also found in the Rocky Mountains.

Edibility: Not recommended.

Note: This is one of several mushrooms that mimic the candy cap, but the mild odor and sticky (or at least shiny) cap distinguish it. *L. subflammeus* is similar but has a slightly peppery taste. In the southern Rockies, *L. alpinus* is abundant; it has a dry, orange to red-brown cap and mild odor. See MD 79 and 82 for more information.

Yellow-staining Milk Cap
(*Lactarius xanthogalactus*)

Other Names: *Lactarius vinaceorufescens.*

Key Features:
1. Cap pinkish to cinnamon or reddish-brown, bald.
2. Gills white or tinged cap color, exuding a milk when cut.
3. Milk white, but quickly turning yellow when exposed.
4. Stalk colored like cap or paler, *without* large spots or pits.
5. Entire mushroom brittle, the stalk snapping open cleanly like a piece of chalk.
6. Flesh staining yellow when cut.
7. Veil, ring, and volva absent.

Other Features: Medium-sized; cap often faintly zoned or spotted; gills attached to stalk; taste mild, bitter, or peppery; spores white to yellowish.

Where: On ground under both hardwoods and conifers, often in large numbers; abundant in California, but exact distribution uncertain.

Edibility: Not recommended; at least some yellow-staining milk caps are poisonous.

Note: The yellow-staining milk and reddish cap are the principal features of this mushroom. It frequently mingles with edible look-alikes such as the bleeding milk cap and candy cap. See MD 74–75 (as *L. vinaceorufescens*) for more information.

Black and White Russula
(*Russula albonigra*)

Other Names: Blackening Russula, Integrated Russula.

Key Features:
1. Cap and stalk white when young, but soon developing gray or black stains and eventually entirely black.
2. Gills white when young but blackening in old age or where bruised.
3. Stalk thick (at least ¾″), hard, rigid.
4. Entire mushroom crisp and brittle, the stalk snapping open cleanly like a piece of chalk.
5. Flesh and stalk staining gray or black a few minutes after being bruised or cut, *not* reddening before blackening.
6. Veil, ring, and volva absent.

Other Features: Medium-sized to large, not decaying readily; cap sunken in the center; gills thick, broadly attached to stalk or running down it, usually alternating long and short; spores white.

Where: On ground in woods; widespread and sometimes common, but often overlooked when black.

Edibility: Not recommended; at least one closely related species is poisonous.

Note: This large, hard russula is easily recognized by the blackening of all parts in age or with handling. Entirely white and entirely black specimens are common, but most dramatic are those in which the gills are still white while the cap and stalk have blackened. Several similar russulas stain red or orange-red before darkening. See MD 89–90 for more information.

Short-stemmed Russula
(*Russula brevipes*)

Key Features:
1. Cap with sunken center, white when fresh (but often dirty and/or with buff or tan stains).
2. Gills white or sometimes tinged blue-green, close together, broadly attached to stalk or running down it.
3. Gills *not* exuding a milk when cut or broken.
4. Stalk hard, rigid, >⅝" thick, *without* a bulb at base.
5. Flesh and stalk white, *not* reddening or blackening when bruised.
6. Entire mushroom crisp and brittle, the stalk snapping open cleanly like a piece of chalk.
7. Veil, ring, and volva absent.

Other Features: Medium-sized to large; cap *not* sticky or slimy, the edge rolled under when young; gills usually alternating long and short; stalk often but not always short; taste mild to slightly peppery; spores white or creamy.

Where: Widespread and abundant in woods, hugging the ground or forming "shrumps" (low mounds in the humus).

Edibility: Edible but insipid.

Note: Although attractive when clean and crisp, this harmless, prolific mushroom is constantly maligned because it mimics prized edibles such as the white matsutake and forms promising "shrumps" like those of the king bolete and chanterelle. Its brittleness and blandness have also contributed to a host of unflattering, even hostile characterizations, such as "so large, so mediocre, "better kicked than picked," "better punted than hunted," and "better trampled than sampled." See photo on p. 14, and MD 87–88.

Emetic Russula (*Russula emetica*)

Other Names: Sickener.

Key Features:
1. Cap bright red to pink, *without* scales or warts.
2. Gills white.
3. Stalk white, *without* a bulb at the base.
4. Entire mushroom brittle, the stalk snapping open cleanly like a piece of chalk.
5. Veil, ring, and volva absent.
6. Taste extremely peppery (chew on a small piece of cap, then spit it out).
7. Spores white.

Other Features: Medium-sized; cap at first domed, then flat or with sunken center, the surface sticky when moist; odor mild.

Where: On ground, moss, or rotten wood in forests and bogs, often in large numbers; widespread and common.

Edibility: Not recommended; it causes nausea and vomiting when eaten raw.

Note: This pretty russula is easily recognized by its bright red cap, white stalk and gills, and peppery taste. The "true" *R. emetica* is said to grow with conifers, especially in bogs, but a closely related form is abundant in California under hardwoods. See MD 97–98 for more information.

Rosy Russula (*Russula rosacea*)

Other Names: Rose-red Russula.

Key Features:
1. Cap dark red, bright red, or pinkish-red, *without* scales or warts.
2. Gills white or creamy.
3. Stalk red or pink.
4. Entire mushroom brittle, the stalk snapping open cleanly like a piece of chalk.
5. Veil, ring, and volva absent.
6. Taste peppery (chew on a small piece of cap, then spit it out).
7. Spores creamy or pale yellow.

Other Features: Medium-sized; cap at first domed, then flat or with sunken center; gills attached to stalk; stalk usually ½–1″ thick, *without* a bulb at base; flesh white; odor mild.

Where: In needle duff under conifers (especially pine and spruce) or in grassy areas near conifers, often in troops; widespread, but most abundant on the West Coast.

Edibility: Not recommended; the strong peppery taste is a deterrent.

Note: The white or creamy gills contrast beautifully with the red or rosy cap and stalk. It often grows with another colorful mushroom, the orange and green delicious milk cap (*Lactarius deliciosus*). The peppery taste distinguishes it from the shrimp mushroom. See MD 99–100 for more information.

Shrimp Mushroom
(*Russula xerampelina*)

Other Names: Shrimp Russula, Woodland Russula.

Key Features:
1. Cap red to purple or purple-brown, sometimes partly or entirely green or brown.
2. Surface of cap sticky or slimy when moist, usually with adhering debris when dry.
3. Gills whitish to yellowish or with brownish stains.
4. Stalk with at least a blush of rose (sometimes entirely rosy), staining yellowish when bruised or handled, then brown.
5. Entire mushroom brittle, the stalk snapping open cleanly like a piece of chalk.
6. Taste *not* peppery (chew on a small piece of cap).
7. Odor fishy or shrimpy when old (absent when young).
8. Veil, ring, and volva absent.
9. Spores yellowish.

Other Features: Medium-sized to large; cap broadly domed at first, then flat or with sunken center; gills usually attached to stalk; flesh inside mature stalk spongy and tinged brown.

Where: On ground near or under conifers, often partly hidden by duff or needles; widespread and common.

Edibility: Delicious! The crisp young caps can be sauteed, toasted, or stuffed and broiled; older ones develop a strong shrimpy odor when cooked but are still tasty.

Note: This mushroom is vastly underappreciated, perhaps because it resembles the hordes of other russulas that litter our coniferous forests. Some specimens lack the telltale blush of rose on the stalk; pass these up until you know the species better. See MD 102–103 for more information.

The shrimp mushroom is extremely variable in color. This reddish-capped variety is abundant under spruce in the Rocky Mountains.

Take Two

Take two nice flat caps. Remove the stems. Square them off. Then slip them into a pop-up toaster. It'll start steaming like it's on fire, but don't worry. Two cycles ought to do it.

—*Jim Arnold*

(Warning: Mushrooms conduct electricity!)

My Most Memorable Mushroom Hunt

We were somewhere left of Boletus Boulevard. As I was walking along I came upon a *very* interesting black umbrella. It had an exquisitely carved handle of some kind of Hindu god. A light rain was starting to fall, so it came in very handy. With great ecstasy I came running up to David and said, "Look at this wonderful umbrella I found!"

And he grabbed it out of my hand and said, "This is my missing Indian umbrella! I lost it out here a year ago when I was hunting!" And so I lost my great find the very second I showed it to him. That was very memorable.

I must add that Rudyard Kipling once said that one of the greatest uses of an umbrella in India is that it is one of the few things that will really protect you from tigers. What you do is aim the point at the tiger and then open and close it very quickly. The tiger thinks it's a puff adder.

—*Bo Hinrichs*

Fragrant Russula
(*Russula fragrantissima* & *R. laurocerasi*)

Other Names: Almond-scented Russula.

Key Features:
1. Cap yellow-brown to ochre, tan, or straw-colored.
2. Edge of cap with furrowed bumps (bumpy furrows).
3. Gills white to yellowish (but often spotted with brown).
4. Odor strongly fragrant (like maraschino cherries or almond extract), sometimes with an unpleasant component also.
5. Stalk at least ½" thick, white when fresh but often yellowish or brownish-spotted in age.
6. Entire mushroom brittle, the stalk snapping open cleanly like a piece of chalk.
7. Veil, ring, and volva absent.

Other Features: Medium-sized to fairly large; cap surface sticky or slimy when moist; gills often beaded with water droplets when fresh; taste unpleasant; spores pale orange-yellow.

Where: Alone or in groups on ground in woods and at their edges; widespread and common.

Edibility: Poisonous.

Note: The thick, heavy aroma of maraschino cherries is unmistakable. The two closely related species described here differ microscopically. See MD 92–93 for more information.

Above Left: A large oyster mushroom (pp. 34–35). **Above Right:** A "boysterous" mushroom hunter. **Below:** The man on horseback (p. 61).

3 Miscellaneous Light-spored Gilled Mushrooms without a Ring

These mushrooms have white to buff, yellowish, pinkish, flesh-colored, or pale lilac spores. The stalk does not snap open cleanly as in the milk caps and russulas (in other words, it contains at least some fibers). The gills are thin-edged and well developed (not blunt as in the chanterelles), and usually pale or brightly colored. In *Hygrophorus* there is a transient veil that occasionally forms a slight ring or volva on the stalk. In the other mushrooms depicted here, a veil, ring, and volva are completely lacking.

Only a small sampling of mushrooms with the above features are shown here, beginning with those that grow on wood. For a more comprehensive treatment, see MD 103–189 and 200–258.

Left: This meaty dark-capped oyster mushroom is common on cottonwood and willow. **Right:** This pearly, pale-capped form favors oak.

My Most Memorable Mushroom Hunt

We were looking for oysters, but none of the usual spots were blooming. Then past a clearing we see this tree sticking out of the side of a ravine, with oysters growing out of it 25 feet up, right out over the ravine where we couldn't reach them! To make matters worse, the slope below was thick with brambles, poison oak, and, I'm sure, hundreds of ticks carrying Lyme's disease. It was like one of those animal problems, you know, the dog on the leash wound around the tree — it's so deeply involved that it can't see the situation, but the raccoon can. So we stood back and studied the problem. I had my boletus stick with me for no good reason, and Bo had a large black umbrella because it was going to rain but didn't. So here's what we did: Bo climbed

the tree as far as he could, clinging precariously to the trunk while reaching out to jab the oysters from underneath with my stick, knocking them off one by one. I crouched far below in the brambles, with the open umbrella inverted above my head like a satellite dish, receiving Bo's emissions. And what delicious emissions they were — we got ten pounds of them!

—*William Rubel*

Oyster Mushroom (*Pleurotus ostreatus*)

Other Names: Tree Oyster, Tree Mushroom.

Key Features:
1. Growing shelflike on dead trees, logs, or stumps.
2. Cap medium-sized to large (>2″ broad).
3. Cap bald, white to gray, dark gray, tan, or brown.
4. Gills white or at least pale, running down stalk (if stalk is present).
5. Stalk short, thick, and off-center *or* entirely absent.
6. Veil, ring, and volva absent.
7. Spores whitish or tinged lilac.

Other Features: Flesh fairly thick; gills broad (fairly deep), sometimes yellowish or grayish in age.

Where: In groups or shelving masses on dead hardwoods (or occasionally conifers); widespread and abundant, especially on cottonwood, willow, alder, oak, and orchard trees. It favors cool weather (fall and winter on the coast, spring and fall inland).

Edibility: Delicious grilled or fried; it is very popular and is now grown commercially.

Note: The oyster mushroom is actually a collection of closely related species and ecological forms (see photos on pp. 33–34). Some are pale-capped and fairly thin; others are thick, meaty, and dark-capped. If you carry home an "oyster log" and keep it moist and cool, you should be able to harvest it regularly. See MD 134–135 and plate 27.

Angel Wings (*Pleurocybella porrigens*)

Other Names: *Pleurotus porrigens, Pleurotellus porrigens.*

Key Features:
1. Growing shelflike on rotting conifers.
2. Cap small (1–2½" broad), fan-shaped or tongue-shaped.
3. Cap bald and white when fresh.
4. Stalk absent or present only as a stubby base.
5. Gills thin, white, close together, shallow, running down the stubby base (if present).
6. Veil, ring, and volva absent.
7. Spores white.

Other Features: Cap often cream-colored in age; flesh thin; odor mild.

Where: In colonies on dead conifers, often in large numbers; widespread in northern latitudes. It is especially common in the Pacific Northwest and northern California.

Edibility: Flimsier than the oyster mushroom, and not nearly as flavorful, but popular nevertheless.

Note: The milk-white caps of this mushroom are a beautiful sight in the dark mossy forests where it thrives. It is often mistaken for a white oyster mushroom, but is smaller, thinner, and favors conifers. See MD 135–136 (as *Pleurotus porrigens*) for more information.

> Mushroom hunting;
> They don't run away,
> But everyone's in such a hurry!
> —Anonymous Japanese

Velvet Foot (*Flammulina velutipes*)

Other Names: Velvet-stemmed Flammulina, Winter Mushroom.

Key Features:
1. Cap fairly small (usually <2" broad), red-brown to orange-brown to bright yellow-orange.
2. Cap broadly rounded to flat or wavy, *not* cone-shaped.
3. Surface of cap bald and sticky or slimy when moist.
4. Gills white to pale yellow.
5. Stalk slender, tough, becoming velvety and rusty-brown to brownish-black from the base upward as it matures (i.e., mature stalk dark and velvety except at very top, young stalk paler and less velvety).
6. Veil, ring, and volva absent.
7. Spores white.
8. Growing in tufts or clusters on dead wood.

Other Features: Gills typically attached to stalk but often barely; stalk of more or less equal width throughout.

Where: Scattered, tufted, or densely clumped on dead hardwoods (especially aspen, willow, poplar, and bush lupine); widespread and common. It often fruits during winter thaws but can be found at almost any time.

Edibility: Bland, but can be eaten when more flavorful fungi are unavailable. The sticky skin should be removed prior to cooking.

Note: Be sure not to confuse this mushroom with brown-spored wood-rotters such as the deadly galerina. A cultivated version, the enokitake (enoki, for short) is popular with the Japanese. It is pure white and beansproutlike with a minuscule cap and almost no taste. See MD 220 for more information.

Deer Mushroom (*Pluteus cervinus*)

Other Names: Fawn Mushroom, *Pluteus atricapillus.*

Key Features:
1. Cap dark brown to tan, bald.
2. Gills free from stalk and very close together.
3. Gills white when young but pinkish or flesh-colored at maturity.
4. Stalk *without* bluish or greenish stains at base.
5. Veil, ring, and volva absent.
6. Spores pinkish or flesh-colored.
7. Growing on wood.

Other Features: Medium-sized; cap at first domed, then becoming flat; flesh often soft; edges of gills *not* brown or black; stalk white or with brownish fibers; odor usually radishlike.

Where: Widespread and common on wood or wood chips, often in groups but not usually clustered.

Edibility: Edible; the flavor varies from good to bad, the texture from soft to flimsy.

Note: This common mushroom has several close relatives, which are best differentiated microscopically; this may account for some of the apparent variation in flavor. To avoid confusion with poisonous species of *Entoloma*, eat only those with unequivocally free gills that are obviously growing on wood. See MD 255–256 for more information.

Giant Sawtooth
(Neolentinus ponderosus)

Other Names: Ponderous Lentinus, *Lentinus ponderosus.*

Key Features:
1. Cap medium-sized to very large (3–20" broad).
2. Cap white to yellowish or tan, with prominent scales.
3. Gills white to yellowish, decurrent (running down stalk).
4. Edges of gills serrated when mature (i.e., finely toothed like a saw).
5. Stalk thick, tough, central to slightly off-center.
6. Veil, ring, and volva absent.
7. Spores white.
8. Growing on dead conifers.

Other Features: Often staining yellowish or orangish with age or handling; flesh thick, white; odor mild or fragrant.

Where: Solitary or in small groups on or near dead mountain conifers (especially pine); found in most mountain ranges, but especially common in the Sierra Nevada from late spring through early fall.

Edibility: Popular in California with Asian-Americans as a matsutake and shiitake substitute. The tough flesh requires thorough cooking.

Note: This mushroom may arise from buried wood, but can still be recognized by its large size, scaly cap, and decurrent, white or yellowish gills with serrated edges. A similar woodrotter, *N. lepideus* (=*Lentinus lepideus*), is also scaly but somewhat smaller, with a veil when young. See MD 142-143 for more information.

Western Jack O' Lantern Mushroom
(*Omphalotus olivascens*)

Other Names: Jack O' Lantern Mushroom.

Key Features:
1. Entirely yellow-orange to olive-yellow when fresh.
2. Cap bald, *not* sticky or slimy.
3. Gills running down the stalk, thin-edged and broad (deep), *not* repeatedly forked or connected by cross-veins.
4. Flesh in cap and stalk colored more or less like the surface, *never* white.
5. Veil, ring, and volva absent.
6. Spores white.
7. Growing on or near hardwood stumps and trees (may appear to be on ground if originating from roots).

Other Features: Medium-sized to fairly large; gills glow in the dark when producing spores (see photo on facing page).

Where: In groups or clumps on stumps or near the bases of chinquapin, oak, madrone, and other hardwoods. Abundant in southern and central California, extending into Oregon but less frequent there.

Edibility: Poisonous; it causes profuse sweating, gastrointestinal distress, and other symptoms of muscarine poisoning (see MD 894).

Note: There is no good reason why this mushroom should be mistaken for a chanterelle, but it sometimes is (see facing page). Note that the flesh is never white. To see the gills glow, sit in a dark closet with a fresh specimen and a grilled cheese sandwich. As you nibble away (on the sandwich, not the mushroom!), your eyes will gradually adjust to the darkness until you can see each gill eeerily outlined. See MD 146–148 and plate 40.

The western jack o' lantern mushroom dyes wool purple, as well as green or blue-green depending on the mordant used.

My First Mushroom Hunt

It was 25 years ago. There were plenty of mushrooms but no good mushroom books for California. People used to say things like if it peeled easily or if it grew on trees it was all right. Well, I picked a yellowish-orange mushroom and since it grew near an oak I thought it was a chanterelle. I cooked it for dinner, and spent the next couple days with both ends open, delirious. It was a jack o' lantern that I ate, and I'm glad I did because it humbled me. It taught me to be careful, to be discerning, to learn more. Now I eat chanterelles all the time, and there's *no way* I could confuse one with a jack o' lantern.
—*Simon Kelly*

Glowing *Omphalotus* gills.

Plums and Custard
(*Tricholomopsis rutilans*)

Other Names: Red-tufted Wood Tricholoma.

Key Features:
1. Cap and stalk with purple-red to red scales or fibers on a yellow background (the scales or fibers usually sparser in age).
2. Gills pale yellow to yellow, attached to stalk but *not* running down it.
3. Flesh pale yellow.
4. Veil, ring, and volva absent.
5. Spores white.
6. Growing on rotten wood.

Other Features: Medium-sized; cap domed when young, becoming flatter with age.

Where: Alone or in small groups on or near rotting conifers, conifer stumps, or wood chips; widespread and fairly common.

Edibility: Mediocre.

Note: The beautiful combination of red or purple-red scales on a yellow background is distinctive. Several species of *Gymnopilus* are similar but have rusty-orange spores. A related white-spored mushroom, *T. decora*, has a yellow cap with small gray to black scales or a gray-black center. It also grows on rotting conifers. See MD 145–146 for more information.

False Chanterelle
(*Hygrophoropsis aurantiaca*)

Other Names: *Clitocybe aurantiaca.*

Key Features:
1. Mature cap flat to shallowly concave, *not* sticky or slimy.
2. Gills orange to deep orange, running down the stalk, close together and thin-edged at maturity.
3. Each gill forked one to three times, but *not* connected to each other by cross-veins.
4. Stalk usually less than ½" thick (occasionally thicker), *not* snapping open cleanly like a piece of chalk.
5. Veil, ring, and volva absent.
6. Odor mild.
7. Spores white or pale cream.
8. Associated with conifers.

Other Features: Small to medium-sized; cap color variable, but usually orange to dark brown, or often brown toward center and orange to yellow-orange at edge; flesh thin, pale or tinged cap color; stalk orangish or brownish, usually well developed.

Where: On ground or rotten wood near or under conifers; widespread and common. It favors cool weather and is sometimes abundant when other fungi are scarce.

Edibility: Not recommended (possibly poisonous).

Note: This variable mushroom can be distinguished from the true chanterelle by its thinner, brighter orange, more crowded gills and by its often browner cap and thinner flesh (although variants with pale orange gills or a whitish cap also occur). The forked gills and growth with conifers distinguish it from the western jack o' lantern mushroom. See MD 479–480 and plate 110.

Witch's Hat (*Hygrocybe conica*)

Other Names: Conical Waxy Cap, Blackening Waxy Cap, *Hygrophorus conicus.*

Key Features:
1. Cap distinctly pointed or cone-shaped.
2. Cap bright yellow, orange, or red when fresh (but may blacken with age).
3. All parts of mushroom (especially the stalk) blackening when handled or with age.
4. Veil, ring, and volva absent.
5. Spores white.

Other Features: Cap dry or sticky, bald, sometimes greenish-yellow; gills whitish to gray or yellow, fairly thick and slightly waxy to the touch; stalk at first whitish to yellow, orange, or reddish, often long and slender, easily splitting lengthwise.

Where: On ground or moss in woods or near trees, or sometimes in grassy areas; widespread and common.

Edibility: Not recommended.

Note: The blackening of this white-spored mushroom has been confused with the blue staining of the dark-spored, hallucinogenic psilocybes. Several similar species of *Hygrocybe* do not blacken when handled. See MD 115–117 and plate 19.

Scarlet Waxy Cap (*Hygrocybe punicea*)

Other Names: *Hygrophorus puniceus.*

Key Features:
1. Cap bright red to orange-red when fresh, sticky or slimy when moist.
2. Gills thick, well spaced, soft and waxy to the touch.
3. Gills yellow to peachy or reddish, free from stalk or attached to it but *not* running down it.
4. Stalk yellow, or sometimes flushed red-orange, the base usually whitish.
5. Stalk usually at least ⅜" thick.
6. Veil, ring, and volva absent.

Other Features: Fairly small to medium-sized; cap bald, domed when young becoming flatter in age; stalk fibrous, *not* snapping open cleanly like a piece of chalk; spores white.

Where: In forest humus, along creeks, and in other damp places; widespread, but especially common during cold weather in the coastal forests of central and northern California. It favors habitats where other mushrooms are scarce (e.g., under redwood and bay laurel).

Edibility: Poisonous to many people.

Note: The thick, waxy gills and brilliant color separate this mushroom from all but a few of its relatives. *Hygrocybe coccinea* (MD plate 20) is similar but has a drier, smaller cap and a redder, slimmer, yellow-based stalk. *Hygrophorus speciosus* (MD plate 15) has gills that run down the stalk and grows with larch or pine. See MD 114–115, 126, and plate 21.

Parrot Mushroom
(*Hygrocybe psittacina*)

Other Names: Parrot Waxy Cap, *Hygrophorus psittacinus.*

Key Features:
1. Small (cap up to 1¼″ broad, stalk <¼″ thick).
2. Cap partly or entirely bright green, dark green, or olive-green when fresh (but soon fading).
3. Both cap and stalk slimy when moist.
4. Gills *not* white.

Other Features: Cap bald, soon fading to some shade of yellow, orange, or pink; gills soft and slightly waxy to the touch, attached to stalk but *not* running down it; gills and stalk variously colored (usually greenish when young but fading); odor mild; veil, ring, and volva absent; spores white.

Where: On ground or moss in woods and bogs, along trails, etc.; widespread. It is especially common in the redwood forests of northern California, where the photograph was taken, but is easily overlooked.

Edibility: Too small and slippery to bother with.

Note: This is our only small green gilled mushroom with a slimy cap and stalk. When the green color has faded it is easily confused with other species. See MD 118–119 for more information.

Cowboy's Handkerchief
(*Hygrophorus eburneus*)

Other Names: Ivory Waxy Cap.

Key Features:
1. Entire mushroom white.
2. Both cap and stalk covered with a thick layer of slime when moist.
3. Gills broadly attached to stalk or running down it, soft to the touch.
4. Stalk usually rather slender (<½" thick), of equal width throughout or tapered toward base.
5. Volva absent (no sack or cup at base of stalk).
6. Spores white.

Other Features: Small to medium-sized; gills at first covered by a transient layer of slime; stalk *not* snapping open cleanly like a piece of chalk; ring absent.

Where. On ground in woods and at their edges, often in groups; widespread. It is especially abundant in the oak-madrone woodlands of California.

Edibility: Too slippery to be of value.

Note: Few mushrooms are as slimy as this one, hence the popular name "cowboy's handkerchief." *H. gliocyclus* is just as slimy but has a cream-colored cap, thicker stalk, and grows with pine. Several other white waxy caps have a dry, not slimy, stalk. See MD 119–122 for more information.

White Alpine Waxy Cap
(*Hygrophorus subalpinus*)

Other Names: Subalpine Waxy Cap.

Key Features:
1. Entire mushroom white or creamy-white.
2. Cap broadly domed when young, flat or wavy in age.
3. Gills decurrent (running down the stalk), rather soft, *not* prominently forked or veined.
4. Stalk >½" thick at top, stout, *not* sticky or slimy, the base often with a bulb when young.
5. Stalk *not* snapping open cleanly like a piece of chalk (i.e., at least somewhat fibrous).
6. Odor mild (even when flesh is crushed).
7. Veil present when very young, soon disappearing or forming a slight ring at or below middle of stalk.
8. Spores white.
9. Growing on ground under mountain conifers in spring, often near melting snow.

Other Features: Medium-sized; cap surface bald, slightly sticky when wet; gill edges *not* serrated; ring sometimes resembling a volva.

Where: On ground under mountain conifers in spring or early summer (occcasionally fall); widespread and common.

Edibility: Edible but bland; it is often munched on by deer and rodents (as shown in photo).

Note: When not growing near melting snow this mushroom can be difficult to identify. It has many look-alikes that grow during the summer, fall, and winter (the short-stemmed russula, species of *Clitocybe*, other white waxy caps, etc.). When the veil forms a ring on the stalk it might even be mistaken for an amanita, white matsutake, or mock matsutake. See MD 121–122 for more information.

Redwood Rooter
(*Caulorhiza umbonata*)

Other Names: Rooting Redwood Collybia, *Collybia umbonata*.

Key Features:
1. Cap cone-shaped when young, the center remaining pointed when mature.
2. Cap yellow-brown to reddish-brown or tan, bald, *not* sticky or slimy when moist.
3. Gills white, *not* running down the stalk
4. Stalk tapered below the ground to form a long, deep "tap root."
5. Veil, ring, and volva absent.
6. Spores white.
7. Growing under redwood.

Other Features: Medium-sized; stalk stringy and fibrous, splitting easily lengthwise.

Where: Alone or in groups in humus beneath redwoods; common along the coast from central California to southern Oregon.

Edibility: Not recommended.

Note: This is one of the characteristic mushrooms of the redwood forest. It is so firmly anchored by its "tap root" that it resists plucking. If the humus around it is carefully loosened, however, enough of the "root" will come up to facilitate identification. Species of *Phaeocollybia* also have a "tap root," but have brown spores and a sticky cap. See MD 218–219, 413–416, and plate 100 for more information.

A Little Mushroom with a Lot of Flavor

The mere mention of "mushroom hunting" evokes exotic visions of forays into the wild: padding gingerly through the primeval stillness of an old-growth conifer forest; searching for nests of coccoli beneath the bearded boughs of oaks and madrones; rambling leisurely through ravines, fragrant pine woods, and puffball-pocked pastures; or, for the truly hardcore element, clambering through tick-infested manzanita thickets in pursuit of the incomparable matsutake. But a mushroom hunt of a decidedly different order can begin in the familiar confines of your own lawn (assuming you have one) and radiate outward gradually, like the mycelium of a fairy ring mushroom (*Marasmius oreades*).

Veil-less and scale-less, without freckle, speckle, or spine, the delectable fairy ring mushroom is one of suburbia's best-kept secrets; in fact, it is usually regarded as a pest because its mycelium forms circular brown patches in lawns, parks, and cemeteries. A *Marasmius* hunt isn't so much an escape from civilization as an entry into it, a social event (unless you stick exclusively to cemeteries). It means getting to know your neighbors, offering to remove "the blemishes from their premises," encouraging them not to spray pesticides, discussing global warming and other trends or, if yours is a university town, whether Anais Nin symbolically regained her father as Otto Rank suggested. In other words, be prepared to do as much talking as you do walking!

One day when I was picking *Marasmius* on a lawn, a round Polish woman came outside and said my mushrooms were no good. Then she invited me inside, where she proceeded to stuff me with sour cream cookies (she thought I was too skinny). She showed me a picture of her son, who was in the Navy (I think she was lonely). Then she gave me a necklace of dried boletes. They were her last ones, she said, and they came "from the old country, where everything tastes better." They made a fabulous soup. So did the *Marasmius*.

Fairy Ring Mushroom
(*Marasmius oreades*)

Other Names: Scotch Bonnet.

Key Features:
1. Growing in groups or rings in grass.
2. Cap small (<2" broad and usually ±1"), often (but not always!) with a large blunt knob at center.
3. Cap whitish to tan, bald, *not* sticky.
4. Gills white to pale tan, fairly well spaced and broad (deep), *not* running down the stalk.
5. Stalk thin (±⅛"), tough, pliant.
6. Veil, ring, and volva absent.
7. Spores white.

Other Features: Dried specimens revive when moistened; gills free from stalk or attached to it; flesh white; odor like cyanide (peach pits) if many are stored together in a bag; stalk base *without* a bulb.

Where: In groups, rings, or arcs in lawns and other grassy places; widespread and abundant, especially along the West Coast. It fruits year-round except during cold weather.

Edibility: Excellent. The stems are tough so use only the caps (you can keep them clean by harvesting with scissors). Sun-dried specimens revive beautifully when moistened.

Note: This nondescript little mushroom makes up in abundance what it lacks in substance. Since many other little brown mushrooms grow on lawns, be sure you identify it correctly before eating it. The spores are white, the cap is often knobbed, and the gills do not run down the stalk. See MD 208–209 for more information.

Sweat-producing Clitocybe
(*Clitocybe dealbata*)

Other Names: Sweating Mushroom, *Clitocybe sudorifica*.

Key Features:
1. Cap small (<2" broad), flat or concave when mature (*without* a central knob).
2. Cap grayish to dull white, *not* sticky or slimy.
3. Gills broadly attached to stalk and usually running down it slightly, close together, pale.
4. Odor *not* fragrant or anise-like.
5. Veil, ring, and volva absent.
6. Spores white.
7. Growing in grass, often in groups but *not* in large clusters.

Other Features: Stalk slender, colored like cap or whiter.

Where: In pastures and other grassy areas, usually in groups or rings; widespread and fairly common.

Edibility: Poisonous. The toxin is muscarine, which causes profuse sweating, vomiting, diarrhea, etc. (see MD 894).

Note: This little lawn lover is sometimes mistaken for the edible fairy ring mushroom, but its gills are closer together and extend down the stalk. Because of its small size and humble appearance it is seldom eaten except by grazing children. It has many look-alikes that grow in the woods. See MD 163 for more information.

> *Mushroom hunting;*
> *Someone not good at it,*
> *With an armful of wildflowers.*
> *—Issa*

Blue-green Anise Mushroom
(*Clitocybe odora*)

Other Names: Anise-scented Clitocybe.

Key Features:
1. Cap *and/or* gills tinged blue-green.
2. Gills broadly attached to stalk or running down it.
3. Odor fragrant, like anise or licorice.
4. Veil, ring, and volva absent.
5. Spores buff, cream, or pinkish-buff.

Other Features: Medium-sized to fairly small; surface of cap *not* sticky or slimy; gills white to buff in one form, bluish-green or greenish in another; stalk colored like cap or gills.

Where: On ground under northern conifers (mainly spruce and fir) and under oak; common in the Rocky Mountains, Southwest, and Pacific Northwest, rare in California.

Edibility: Can be used as a flavoring in cookies or breads; it tastes like it smells.

Note: Several species of *Clitocybe* have a strong anise-like odor, and several have a bluish or greenish cap, but this is the only one with both features. *C. fragrans* and *C. deceptiva* (shown below) are two small, whitish to tan, anise-scented species. They are also edible but closely resemble poisonous types. See MD 161–163 for more information.

Blewit (*Clitocybe nuda*)

Other Names: Wood Blewit, *Lepista nuda, Tricholoma nudum*.

Key Features:
1. Entire mushroom purple or bluish-purple, or at least purple-tinged when fresh (but fading, especially the cap).
2. Cap bald, *not* sticky or slimy, the edge tucked under when young.
3. Gills close together, attached to stalk.
4. Stalk at least ½" thick, *without* scales or hairs.
5. Stalk solid (*not* hollow).
6. Veil, ring, and volva absent.
7. Spores pale (pinkish or cream).

Other Features: Medium-sized; cap often shiny when dry, brownish to gray, tan, or paler when not purple; flesh purple or lilac-tinged; gills thin, fading to pinkish-tan or brownish; base of stalk often swollen and often with purple fuzz.

Where: Widespread and ubiquitous in humus or debris in woods, gardens, landscaped areas, etc., often in rings. It is especially common on the West Coast.

Edibility: Popular, although bitter specimens are sometimes encountered (especially under cypress); it should be cooked.

Note: The amount of purple present depends on age, environment, and other factors. Beginners should eat only those that are unequivocally purple; blewits showing just a tinge of purple are easily confused with poisonous species. Other purple mushrooms include *Inocybe lilacina* (poisonous!) and many *Cortinarius* species (see photo at bottom of facing page), with brown spores; *Mycena pura*, with a hollow stem; and the western amethyst laccaria, with a slimmer, hairier stalk and thicker, sparser gills. See MD 153–154 and plate 32.

How to Do It to the Blewit

Blewits can add color and flavor to any dish, but can also be watery or rubbery. The key is in the preparation. A *dry-saute* is best. Clean and slice the mushrooms, then put them in a skillet on *maximum* heat *with no butter or oil.* Sprinkle them with salt (to help draw out the moisture). Stir constantly until they begin to give off liquid, then let them cook in their own juices. After the liquid has evaporated (you can pour off some of it for use as stock), continue to cook them for a couple minutes, stirring constantly again so they don't stick. Then turn down the heat, add a little olive oil or butter, and cook for another 5-10 minutes. They should be fairly chewy and full of flavor, not slimy. You can eat them as they are or incorporate them into any dish. The dry-saute method is also ideal for boletes, chanterelles, hedgehogs, and other mushrooms with a high water content.

Blewits (two at center) with two purple *Cortinarius* of unknown edibility. The blewit lacks a veil and has pinkish spores, while *Cortinarius* has a cobwebby veil when young and rusty-brown spores.

Cloudy Clitocybe (*Clitocybe nebularis*)

Other Names: Skunk Mushroom, Graycap, *Lepista nebularis*.

Key Features:
1. Cap gray to grayish-brown or grayish-tan, at least 3″ broad when mature.
2. Surface of cap *not* sticky or slimy.
3. Gills whitish to buff, close together, decurrent (running down the stalk).
4. Stalk >½″ thick, *not* snapping open cleanly like a piece of chalk (i.e., at least somewhat fibrous).
5. Veil, ring, and volva absent.
6. Odor unpleasant (like skunk cabbage or mice cages), especially when flesh is crushed.
7. Spores buff or pale yellowish.

Other Features: Medium-sized to large; stalk white or colored like cap or gills, the base often enlarged but *without* a large mat of white threads (mycelium) attached.

Where: On ground in woods and at their edges, usually in groups or large rings; widespread and common, especially during cool weather.

Edibility: Not recommended.

Note: This large grayish mushroom is best recognized by its obnoxious odor, decurrent gills, and buff-colored spores. *C. robusta* has the same odor and spore color but has a whitish cap. Other similar mushrooms lack the odor or have white spores. See MD 159–160 for more information.

Western Amethyst Laccaria
(*Laccaria amethysteo-occidentalis*)

Key Features:
1. Entire mushroom amethyst-purple when fresh and moist.
2. Cap usually <2" broad.
3. Cap and stalk soon fading to brown, tan, or paler.
4. Gills fairly thick and well spaced, fading like the cap but more slowly.
5. Stalk slender (<½"), of more or less equal width throughout, *without* a bulb at base.
6. Stalk often long; tough and fibrous, usually with hairs or fibers running lengthwise.
7. Veil, ring, and volva absent.
8. Spores white or tinged lilac.

Other Features: Not decaying quickly; flesh violet or tinged violet; odor mild; base of stalk often with purple fuzz.

Where: On ground in woods and chaparral and near trees (especially conifers but also oak); abundant on the West Coast, possibly more widespread.

Edibility: The caps are bland but have a nice, chewy texture; they can be sauteed whole and flavored with garlic or onion.

Note: The striking amethyst color is especially evident in the gills of young specimens. Many *Cortinarius* species are purple but have rusty-brown spores, while the blewit has a stouter, thicker, smoother stalk and pinkish-cream spores. Several other species of *Laccaria* are pinkish or flesh-colored. See MD 171–173 and plates 29–30.

Large White Leucopaxillus
(*Leucopaxillus albissimus*)

Other Names: White False Paxillus.

Key Features:
1. Cap white when young (but often aging yellowish).
2. Surface of cap dry and dull (unpolished).
3. Gills white or whitish, running down the stalk.
4. Stalk white or slightly yellowish, *not* snapping open cleanly like a piece of chalk (i.e., fibrous).
5. Base of stalk with a mat of white threads that may or may not come up with the mushroom when plucked.
6. Veil, ring, and volva absent.
7. Entire mushroom rather tough, persisting for days or even weeks without decaying.
8. Spores white.
9. Growing on ground.

Other Features: Medium-sized to very large; cap bald or cracked; gills close together, their edges typically *not* serrated; stalk often curved or bent and tapered at base; odor distinctive but difficult to describe; taste mild or bitter.

Where: Widespread on ground in woods, often in groups or rings; especially commmon along the coast under redwood and spruce.

Edibility: Not recommended.

Note: Several varieties of this mushroom have been described based on differences in color, size, and taste. Its resistance to decay has been attributed to the presence of antibiotics. In the Southwest, another large whitish mushroom, *Clitocybe candida*, is more common. It lacks the mat of white threads. See MD 158–159 and 167–168 for more information.

Bitter Brown Leucopaxillus
(*Leucopaxillus amarus*)

Other Names: Bitter False Paxillus, *Leucopaxillus gentianeus*.

Key Features:
1. Cap brown to reddish-brown, bald.
2. Surface of cap dry and dull (unpolished).
3. Gills white, close together, attached to stalk.
4. Stalk white, >⅜" thick, *not* snapping open cleanly like a piece of chalk (i.e., at least somewhat fibrous).
5. Base of stalk with a mat of white threads that extends into the humus and usually comes up with the mushroom when plucked.
6. Taste very bitter (chew on a small piece of cap, then spit it out).
7. Veil, ring, and volva absent.
8. Spores white.
9. Growing on ground.

Other Features: Medium-sized; gills attached to stalk and often notched where they join it.

Where: On ground in woods and under trees (especially oak, pine, and other conifers), often in rings; widespread and common.

Edibility: Too bitter to eat.

Note: This boring brown-capped mushroom resembles many others, but a telltale mat of moldy-looking white threads (the mycelium) comes up with the stalk if it is plucked carefully. See MD 168 for more information.

Poplar Tricholoma
(*Tricholoma populinum*)

Other Names: Sandy.

Key Features:
1. Growing in masses under poplar (cottonwood).
2. Cap pinkish-brown or reddish-brown toward the center, the edge usually paler.
3. Gills attached narrowly to stalk, *not* running down it.
4. Gills and stalk white when fresh but developing reddish-brown stains with age or where bruised.
5. Stalk >⅜" thick, *not* snapping open cleanly like a piece of chalk (i.e., at least somewhat fibrous).
6. Veil, ring, and volva absent.
7. Odor of crushed flesh like meal or cucumber.
8. Spores white.

Other Features: Medium-sized; cap bald, broadly domed at first, flat or wavy in age; gills close together; stalk *without* scales or a mat of white threads at base.

Where: Widespread in sand or soil under poplar, usually along rivers or streams; especially common inland, where it is one of the last mushrooms to fruit in the fall. It typically develops underground, then surfaces in rings or densely packed masses of up to a thousand individuals.

Edibility: Popular in the Great Basin and prairie regions where poplars are prevalent and other trees scarce. It has been eaten for centuries by Native Americans from British Columbia to the Southwest, and is still collected every fall by residents of the Taos Pueblo in New Mexico.

Note: The habitat is a key feature of this mushroom. Several very similar, poisonous tricholomas (see MD plate 35) grow with other trees, whereas few white-spored mushrooms occur in masses in sandy soil under poplar. See MD 185–187 for more information.

Man On Horseback
(*Tricholoma flavovirens*)

Other Names: Canary Trich, *Tricholoma equestre*.

Key Features:
1. Gills evenly yellow and notched (dipping in) where they join the stalk.
2. Cap entirely yellow, or brown to reddish-brown with a yellow edge.
3. Surface of cap bald and sticky when moist, usually with adhering debris when dry.
4. Stalk white or pale yellow, *without* scales.
5. Stalk >⅜" thick, *not* snapping open cleanly like a piece of chalk (i.e., at least somewhat fibrous).
6. Veil, ring, and volva absent.
7. Flesh white.
8. Spores white.

Other Features: Medium-sized; cap at first broadly domed with the edge tucked under, eventually flat or wavy; odor pleasant.

Where: On ground in woods, usually in groups but often visible only as "shrumps" (low humps) in the humus; widespread and common in cool weather under pine, spruce, aspen, and madrone.

Edibility: Delicious, but greatly underappreciated.

Note: This mushroom is as appealingly yellow as the blewit is purple. Uncovering some in deep pine needle duff can be like digging up nuggets of gold. A related mushroom of uncertain edibility, *T. sejunctum*, shows less yellow in the gills and has dark fibers at the center of the cap. *T. sulphureum* is also yellow but has yellow flesh and smells obnoxious. See p. 33 for another photo, and MD 179–180.

Fried Chicken Mushroom
(*Lyophyllum decastes*)

Other Names: Clustered Lyophyllum.

Key Features:
1. Growing in large clumps in disturbed ground, the individual stalks joined at the base.
2. Cap dark brown, brown, or tan, bald.
3. Gills white.
4. Stalk white, usually curved, at least ½" thick.
5. Veil, ring, and volva absent.
6. Spores white.

Other Features: Medium-sized; cap *not* sticky when moist but may have a slightly soapy feel; gills attached to stalk; odor mild.

Where: Widespread and often abundant in clumps of up to a hundred individuals, usually in disturbed areas (along roads and paths, in waste places, around old sawdust piles, etc.). The clumps are often half hidden by leaves or grass.

Edibility: Edible. In Japan, several species of *Lyophyllum* are marketed under the name *shimeji*. North American variants are plentiful but not as flavorful, and occasionally cause stomach upsets.

Note: No one seems to know how this mushroom got its name — it certainly doesn't taste anything like chicken! The dark-capped variety is often regarded as a distinct species, *L. loricatum*. Be sure not to eat clumped mushrooms with white to gray caps or with pinkish spores, as they may be poisonous. See MD 174–175 for more information.

Above: An amanita "egg" sliced open vertically to show the developing cap, gills, and stalk encased within the universal veil. **Below:** A spring coccora (p. 69) with gills still hidden by the partial veil.

4 Amanita

This group contains the most deadly of all mushrooms (pp. 64–66) as well as some of the most beautiful and delicious. All amanitas have pale gills (usually white or yellow) and white spores, and are initially surrounded by a universal veil. As the mushroom emerges, the universal veil ruptures, typically forming a volva (sack, cup, collar, or series of scaly rings) at the base of the stalk, and often leaving remnants on the cap (a large patch or many small scattered flakes or warts). Many amanitas also have a partial veil that shrouds the young gills and, upon rupturing, often forms a ring near the top of the stalk. The gills can be free from the stalk or attached to it, but do *not* run down it.

About half of the common western amanitas are depicted here, plus one look-alike with pinkish spores, *Volvariella*. If your "amanita" is not among these, check the shaggy stem (p. 90) and shaggy parasol (p. 95), and consult MD 258–290.

Death Cap (*Amanita phalloides*)

Key Features:
1. Cap pale green to green, greenish-yellow, or olive-brown when fresh (but often fading to tan or paler).
2. Cap usually bald (*without* warts, but sometimes with a thin white patch of universal veil tissue).
3. Edge of cap typically *without* radial lines or furrows.
4. Gills white in all stages.
5. Partial veil present, at first covering the gills, then usually forming a skirtlike ring on upper stalk.
6. Universal veil white, enveloping the young mushroom, then forming a sack or cup (volva) at base of stalk.
7. Always growing near or under trees.

Other Features: Medium-sized; cap often with a metallic luster; odor usually pungent in age; stalk white or tinged green, base usually thickened or with a bulb; spores white.

Where: On ground in woods and near planted trees, often in groups; abundant throughout much of California under oak, and recently found in Oregon and Washington. It was probably introduced from Europe, and appears to be spreading.

Edibility: Deadly poisonous! A single mushroom can be fatal if eaten, but there is no harm in handling it.

Note: Some death caps may lack one or more of the above features. For instance, rain or sun can bleach the cap or cause small furrows to form on its edge, the ring may disappear, or the volva can be carelessly overlooked. If you value your life, learn to recognize the deadly amanitas in *all* their guises before eating any white-gilled mushroom. See facing page and MD 269–271, 892–893, and plate 50.

Death cap (*Amanita phalloides*). Beware of specimens with little or no green in the cap; they may not look like the classical ones shown on facing page, but can still be recognized by the features discussed below.

Destroying angel (*Amanita ocreata*, p. 66).

Veiled Threats: The Deadly Duo

The death cap (facing page) and destroying angels (next page) are the most poisonous mushrooms known. Fortunately, they are fairly easy to recognize if dug up carefully. Their telltale features are the white gills (and white spores) and presence of a sack (volva) at the base of the stalk (the remnants of the universal veil). In addition, a partial veil shrouds the young gills and often forms a fragile ring on the upper stalk (but may shred into many pieces as shown in above photo), the cap is usually bald (but sometimes has a thin patch of universal veil tissue), and the gills do *not* run down the stalk (they may be free or attached).

These amanitas are responsible for nearly all mushroom-induced fatalities. Poisonings are usually the result of erroneous beliefs (for example, the "silver coin" or "red rice" tests) or sheer ignorance rather than misidentification. What makes the "die-namic duo" doubly dangerous is that symptoms do not appear for 6–24 hours after ingestion, by which time the toxins have been completely absorbed.

Destroying Angel (*Amanita ocreata*)

Key Features:

1. Cap pure white, or white at the edge and pinkish, yellowish, tan, or buff at the center.
2. Cap usually bald (*without* warts, but sometimes with a thin white patch of universal veil tissue).
3. Edge of cap usually (not always!) *without* radial lines or furrows.
4. Gills and stalk white in all stages.
5. Partial veil present, at first covering the gills, then disintegrating or forming a fragile skirtlike ring high on the stalk.
6. Universal veil white, enveloping the young mushroom, then forming a sack or cup (volva) at base of stalk.
7. Always growing near trees, especially oak.

Other Features: Medium-sized; gills *not* running down the stalk; stalk slender or stout, usually thickened at base or with a bulb; spores white.

Where: On ground in woods and at their edges; common under oak in California during the winter and spring, possibly in Oregon and the Southwest. In southern California and the Sierra Nevada foothills it is more common than the death cap.

Edibility: Deadly poisonous!

Note: This tantalizing mushroom can be slender or robust; it might be mistaken for the meadow mushroom, horse mushroom, ma'am on motorcycle, spring coccora, or other delectables, yet it has not been implicated in as many poisonings as the death cap. Another deadly, pure white destroying angel, *A. verna*, has been found in the Pacific Northwest. See p. 65 for another photo and more details; also MD 271–273, 892–893, and plate 53.

Southwestern Caesar's Amanita
(*Amanita caesarea*)

Other Names: Caesar's Mushroom.

Key Features:
1. Cap bright red to orange when fresh (but duller or browner after fading).
2. Cap bald or with a large white patch or "skullcap" of universal veil tissue (*without* warts).
3. Edge of cap striate (with many short radial lines or furrows), even when young.
4. Gills bright yellow or sometimes yellow-orange.
5. Stalk yellow, the center hollow (but often filled with a cotton or gel).
6. Partial veil present, at first covering the gills, then usually forming a yellow skirtlike ring on upper stalk.
7. Universal veil white, enveloping the young mushroom, then forming a large, thick cup or sack (volva) at base of stalk.

Other Features: Medium-sized to fairly large; stalk thick, *not* swollen at base; spores white.

Where: On ground in pine forests after summer thunderstorms, often in groups or rings; known only from the Southwest. It is especially common in the Grand Canyon area.

Edibility: Edible; similar to the coccora in flavor and texture.

Note: This is the only western amanita with bright yellow gills and stalk. It is called *A. caesarea* for lack of a better name, but differs somewhat from the true *A. caesarea* of Europe. It is more robust than the slender, bald-capped caesar's amanita of eastern North America (now called *A. hemibapha*). See MD 284 for more information.

Coccora (*Amanita calyptroderma*)

Other Names: Coccoli, Cocconi, *Amanita calyptrata.*

Key Features:
1. Cap dark brown to orange-brown, yellow-brown, or amber (the edge often yellow).
2. Cap with a thick white cottony patch or "skullcap" of universal veil tissue (*without* warts).
3. Edge of cap striate (with many short radial lines or furrows), even when young.
4. Stalk creamy or pale yellow (*not* white), the center hollow but usually filled with a cotton or gel.
5. Partial veil present, at first covering the gills, then usually forming a skirtlike ring on upper stalk.
6. Universal veil white, enveloping the young mushroom, then forming a large, thick cup or sack (volva) at base of stalk.

Other Features: Medium-sized to large; gills whitish to cream; stalk thick, *not* swollen at base; ring creamy or yellowish; spores white.

Where: On ground in woods, often in "families"; California to British Columbia; abundant in central and northern California under oak and madrone after the first fall rains.

Edibility: Very popular with Italian-Americans because of its resemblance to the fabled caesar's amanita of Italy.

Note: To distinguish this beauty from poisonous amanitas, do *not* rely on a single feature such as presence of a "skullcap" or furrowed vs. smooth cap edge. Collect it repeatedly before eating it, and become thoroughly familiar with its poisonous counterparts. The spring coccora (top photo on facing page) requires even more caution. See MD 284–286 and plates 61–63 (as *A. calyptrata*).

The spring coccora (above) resembles the typical coccora (shown below and on facing page) in all respects save color: the cap is dull white to pale yellow and the gills and stalk are whitish. It is also ecologically distinct, growing in late winter and spring along the coast and during the spring morel season in the mountains. Although probably a distinct species, the spring coccora does not yet have its own scientific name. It is not quite as tasty as the fall version, and is more likely to be confused with poisonous amanitas, so beginners beware!

Things Go Better with Coccora

Both the spring coccora (above) and the darker fall coccora (below) are an acquired taste (unless you're Italian-American). If you find them slightly bitter, try sauteeing them in olive oil with shallots and a little minced garlic; then incorporate them into tomato sauce or serve them over pasta or polenta. Unlike many mushrooms, they retain a wonderful crunchiness even after being sauteed and frozen. If you acquire a taste for them, eventually you'll become addicted, and pasta will be barely palatable *without* coccora.

A basket of springtime amanitas.

Boletes or Chanterelles?

Chanterelles are fun to find, but I don't enjoy eating them. I don't know why people think they're so great. Mushroom hunters think I'm in a rut because I like the regular commercial mushroom, but what about *them?* There are these certain mushrooms that are considered choice edibles and everybody goes *flocking* after them. They become cliches. My first experience with chanterelles was in a soup. Since I'd never cooked them before, and since I was *very* busy that week, and since they stayed in the refrigerator too long, it turned out to be — *and it wasn't my fault* — the slimiest, vilest, most disgusting soup I'd ever tasted. A mucilaginous mess with little slugs swimming around in it. Boletes are definitely better: plump, substantial, and delicious when they're not too slimy. *Amanita velosa* is better than both of them.   —*Judith Scott Mattoon*

This ad for Honeywell shows how the deadly poisonous destroying angel (left) can be mistaken for the delicious springtime amanita (right). (Courtesy of Honeywell and Cy DeCosse Inc.)

Looks can be deceiving.

Springtime Amanita (*Amanita velosa*)

Key Features:
1. Cap evenly salmon to pinkish-tan or dull orange, usually with a single white patch of universal veil tissue (but see "Other Features" below).
2. Edge of cap distinctly striate (with many short radial lines or furrows), even when young.
3. Gills white to faintly pinkish.
4. Stalk bluntly pointed at base, *without* a bulb.
5. Partial veil absent; stalk *without* a skirtlike ring.
6. Universal veil white, enveloping the young mushroom, then forming a sack or cup (volva) at base of stalk.
7. Associated with oak, usually in the spring.

Other Features: Medium-sized; cap color variable, ranging to beige or white, sometimes bald or with several large warts at center; stalk white when fresh, slender or stout, sometimes with a scaly ringlike zone; spores white.

Where: On ground near oaks (but often in the open); abundant in southern and central California in the late winter and spring, less frequent northward as far as Oregon.

Edibility: One of the sweetest and most delicious of all mushrooms, but easily confused with the destroying angel and other poisonous species (see photo at bottom of facing page). Your first mistake could be your last!

Note: Don't eat this variable mushroom unless you are thoroughly familiar with it; even then it is wise to reject specimens that do not have an evenly salmon-colored cap adorned with a single patch of veil tissue. A similar delicious species with a slightly oranger, usually bald cap, *A. crocea*, occurs in the Southwest. See MD 286–287 and plate 64.

71

Western Grisette (*Amanita pachycolea*)

Key Features:
1. Cap bald, prominently striate (with long radial grooves extending from the edge toward the center).
2. Cap medium-sized to large (3–10" broad when mature), dark brown to brown or gray, often with a darker band around the inner edge of the furrows.
3. Stalk long (6–20" when mature).
4. Universal veil enveloping the young mushroom, then forming a large sack (volva) at base of stalk.
5. Volva 2–5" high, white or rusty-stained, attached to stalk only at the base.
6. Partial veil absent; stalk *without* a ring.

Other Features: Mature cap sometimes with a broad central hump; gills white to gray or with brown edges; stalk white or often clothed with delicate gray to brown scales; spores white.

Where: Alone or in small groups on ground in woods, from central California to the Pacific Northwest; it is a characteristic feature of the coastal forests.

Edibility: Very tasty, but be sure of your identification.

Note: Grisettes (see facing page) are edible amanitas with a gray to brown cap that is strongly grooved and a volva but no ring on the stalk. This exceedingly handsome species is larger and more flavorful than other grisettes, and has a much more impressive volva. See MD 288–290 and plates 65–66.

Amanita protecta is an oak-loving grisette of central and southern California. It can be recognized by its unusually thick volva, and the tendency of the universal veil remnants on the cap and stalk to turn yellowish with age (not obvious in this photograph).

Other Grisettes

Although very widely distributed, the grisette (*Amanita vaginata*) is not as common on the West Coast as some of the other grisettes. It tends to be slender, with a gray or gray-brown cap that is usually bald. The volva is sacklike, not pinched as in *A. constricta*.

Amanita constricta is common on the West Coast under both hardwoods and conifers. Note how the upper part of its volva flares outward while the lower portion is pinched or constricted around the base of the stalk; the cap color ranges from brown to gray.

Common Volvariella
(*Volvariella speciosa*)

Other Names: Smooth Volvariella, Showy Volvaria.

Key Features:
1. Cap white to grayish, bald.
2. Gills flesh-colored to dark reddish when mature (but white when young).
3. Gills free from stalk.
4. Partial veil absent; stalk *without* a ring.
5. Universal veil enveloping the young mushroom, then forming a small sack or cup (volva) at base of stalk.
6. Spores flesh-colored to dull reddish-brown.

Other Features: Medium-sized; cap sticky when moist; flesh soft, decaying quickly; gills very close together; stalk white.

Where: On ground in gardens and agricultural fields, along roads, in compost, etc.; widespread and common.

Edibility: Excellent when young, but easily confused with poisonous amanitas; older specimens are flaccid.

Note: This mushroom looks like an amanita, but its spores and mature gills are pinkish to dull reddish, not white. In agricultural areas it can fruit so prolifically that the fields look as if they were dusted with snow. See MD 259–260 for more information.

Blusher
(*Amanita rubescens* & *A. novinupta*)

Other Names: Blushing Amanita.

Key Features:
1. Cap white to tan, brown, or dull reddish, with scattered warts or flakes (universal veil remnants).
2. Gills white, free from or attached to stalk but *not* running down it.
3. Stalk with dull reddish stains, especially near base.
4. Flesh slowly reddening when bruised (within several hours); maggot tunnels if present also dull reddish.
5. Partial veil present, at first covering the gills, then forming a skirtlike ring on stalk whose upper surface has gill-like grooves or lines.
6. Universal veil enveloping the young mushroom, *not* forming a volva, or sometimes leaving one or more scaly rings near base of stalk but no distinct cup or sack.

Other Features: Medium-sized; warts on cap dense or sparse; stalk stout or slender, white when young, reddish or brownish in age; spores white.

Where: On ground in woods and at their edges; common in California (especially under oak in the spring), also found in southern Oregon, and in the Southwest under conifers.

Edibility: Edible when cooked, but be certain of your identification! It is said to be poisonous raw.

Note: The blusher shown in the photograph is from California, and has recently been named *A. novinupta*. It differs in serveral respects from the blusher or Europe, eastern North America, and the Southwest, and doesn't taste as good. It stains reddish, however, so for a long time it was called *A. rubescens*. See MD 276-277 for more information.

Fly amanita "caps." For centuries the fly amanita has played a central role in the shamanistic practices of the Far North. It has also permeated our own cultural mythology more than any other mushroom, as evidenced by its frequent appearance in contemporary fashion (above and left), folk art, and kitsch (below). It may even have inspired the attire of that jolly denizen of the Far North, Santa Claus. Or is it just a coincidence that reindeer are notoriously fond of eating fly amanitas, and that Santa's reindeer *fly*?

Fly Amanita (*Amanita muscaria*)

Other Names: Fly Agaric.

Key Features:
1. Cap bright red to red-orange with scattered white or yellow warts or flakes (universal veil remnants).
2. Gills white.
3. Stalk white, with a bulb at the base.
4. Partial veil present, at first covering the gills, then forming a white or yellow-edged skirtlike ring on stalk.
5. Universal veil white or yellow, enveloping the young mushroom, then forming a volva at base of stalk.
6. Volva consisting of one or more scaly rings (*not* a true sack) at the top of the basal bulb.

Other Features: Medium-sized to large; cap often fading with age, the edge usually with short radial grooves or lines; spores white.

Where: Common on ground in woods and at their edges (especially with pine, spruce, and birch) or sometimes under planted trees, often in groups or rings. It is the most widely distributed amanita in North America.

Edibility: Not recommended. Although its intoxicating properties have been known for centuries, it has other potentially dangerous and unpredictable toxic effects.

Note: Few mushrooms are as famous as "the bright red one with white spots." A yellow-spotted version also occurs. A third variety has white spots and a yellow to yellow-orange cap (MD plate 58); it is common in northeastern North America but infrequent in the West. See MD 282–283, 894–895, and plate 59.

Panther Amanita (*Amanita pantherina*)

Key Features:
1. Cap tan to dark brown, with scattered white warts or flakes (universal veil remnants).
2. Gills white.
3. Stalk white, *not* reddening when bruised or in age, with a bulb at the base.
4. Partial veil present, at first covering the gills, then usually forming a skirtlike ring on upper stalk.
5. Universal veil white, enveloping the young mushroom, then usually forming a volva at base of stalk.
6. Volva typically consisting of a free rim or collar at the top of the basal bulb (but sometimes more than one rim, occasionally none).

Other Features: Medium-sized to large; edge of cap usually with short radial grooves or lines; spores white.

Where: Alone or in groups on ground under oak and various conifers; widespread and common.

Edibility: Not recommended. Although deliberately eaten by some people for its consciousness-altering properties, its effects are unpredictable and potentially dangerous. It contains the same toxins as the fly amanita.

Note: This handsome mushroom resembles the fly amanita, but its cap is never red. It seems to intergrade with the gemmed amanita, forming a confusing series of forms with a yellowish-tan or buff-colored cap. See photo on p. *xix*; also MD 280 and 894–895.

Gemmed Amanita (*Amanita gemmata*)

Other Names: Jonquil Amanita, *Amanita junquillea*.

Key Features:
1. Cap yellow to pale yellow, with scattered white flakes or warts (universal veil remnants).
2. Gills white.
3. Stalk white, *not* reddening when bruised or in age, usually with a bulb at base.
4. Partial veil typically present (but absent in one form), covering the gills at first, then often forming a skirtlike ring on upper stalk.
5. Universal veil white, enveloping the young mushroom, then usually forming a volva at base of stalk.
6. Volva sometimes absent but usually distinct as a free rim or collar at the top of the basal bulb.

Other Features: Medium-sized; veil remnants on cap dense or sparse; edge of cap usually with short radial lines or furrows (often faint); spores white.

Where: On ground in woods and at their edges; widespread and fairly common. In the Sierra Nevada and Cascades it fruits during the spring as well as summer and fall.

Edibility: Not recommended; it intergrades with the panther amanita and reportedly contains the same toxins.

Note: Cap color is the main difference between this species and the panther amanita. It is sometimes mistaken for the pale form of the coccora, but does not have a huge sack at the base of the stalk. The yellow-capped variety of the fly amanita is usually more robust, with a larger, scalier bulb and, often, more orange in the cap. See MD 281 and 894–895 for more information.

Yellow-veiled Amanita
(Amanita francheti)

Other Names: Warted Amanita, *Amanita aspera.*

Key Features:
1. Cap bright yellow, yellow-brown, or dark brown, with scattered yellow to grayish warts or powder (universal veil remnants).
2. Gills white.
3. Stalk white or yellow, usually with yellow flakes or powdery rings near base.
4. Partial veil present, covering the gills at first, then usually forming a yellow or partly yellow skirtlike ring on upper stalk.
5. Universal veil yellow when it envelops the young mushroom, usually forming a volva at base of stalk.
6. Volva present only as yellow to grayish-yellow powder or as one or more scaly rings (no distinct sack).

Other Features: Medium-sized; edge of cap typically *without* furrows or lines; stalk often with orangish or reddish stains near base; spores white.

Where: On ground in woods and at their edges; common in the Pacific Northwest and California.

Edibility: Not recommended.

Note: The yellow veil color separates this striking mushroom from the gemmed amanita and panther amanita. In the Pacific Northwest it usually has a dark brown cap, while in California the cap is frequently yellow. The ring is more prominent than in the shaggy stem (*Armillaria straminea*). See MD 278 for more information.

Two delectable collectables: the white matsutake (above) and the shaggy parasol (below). See pp. 86–87 and 94–95 for more details.

5 Miscellaneous Light-spored Gilled Mushrooms with a Ring

In these mushrooms a veil covers the young gills and usually forms a ring on the stalk. There is no volva as in the amanitas, the gills are usually white or yellow, and the spores are white or, in one case, greenish.

The mushrooms in this section are arranged according to gill attachment. First are those whose gills run down the stalk, then those whose gills are attached to the stalk but do not extend down it, and finally those with gills completely free from the stalk (the parasol mushrooms).

This group includes prized edible as well as poisonous species. If your white-spored, ringed mushroom is not depicted here, check the blusher (p. 75) and white alpine waxy cap (p. 48), and see MD 136, 142, 189–200, and 291–310.

Honey Mushroom
(Armillaria mellea)

Other Names: Oak Root Fungus, *Armillariella mellea.*

Key Features:
1. Growing in clumps on trees, logs, and stumps.
2. Gills white to yellowish or flesh-colored but *not* brown, slightly decurrent (running down the stalk).
3. Stalk tough, often long, with a white stringy pith inside.
4. Veil present, at first covering the gills, then forming a ring on upper stalk.
5. Spores white (lowermost caps of a cluster often coated with white spore dust).

Other Features: Medium-sized to large; cap color variable but usually yellowish to brown with scattered darker hairs or tiny scales; taste of raw flesh usually bitter; volva absent.

Where: In clumps on stumps, trees, logs, and roots of many kinds of trees (oaks, orchard trees, conifers, etc.); widespread and abundant.

Edibility: Often eaten, but sometimes causing stomach upsets. Always cook it well and avoid those growing on buckeye or hemlock, as they are more apt to cause trouble.

Note: This extremely common and variable mushroom is actually more than one species. The above set of fieldmarks will not work for the occasional solitary individual or those that seem to be growing on the ground but actually arise from roots or buried wood. Such specimens can confuse even experienced mushroom hunters, and should not be eaten by beginners. Just remember: this is the only clustered white-spored mushroom that grows on wood and has a ring plus stringy white pith in the stalk! See pp. 83-84 for more photos; also, MD 196-197 and plate 42.

Clusters of honey mushrooms. Note how the caps of main cluster in photo on right are coated with white spore dust.

Stalking Up on Honeys

Conventional wisdom is to "discard the tough stalks" of honey mushrooms and use only the caps; as a result, you may occasionally find massive clusters of stalks whose caps have been harvested. Don't pass these up, because the stalks are the best part of the mushroom! Just strip away the outer rind and you are left with a string cheese-like white pith that makes a wonderful appetizer (below) when sauteed in a little olive oil.

Clusters of honey mushrooms (*Armillaria mellea*, p. 82).

Mushroom gathering;
Today let's go on till we fall over
The roots of the trees.
—Kaso

Mock Matsutake
(*Catathelasma ventricosa*)

Other Names: Swollen-stalked Catathelasma.

Key Features:
1. Cap often large (3–15″), white to grayish, *without* prominent scales.
2. Gills white, decurrent (extending down the stalk).
3. Edges of gills *not* serrated.
4. Stalk thick, white, firm.
5. Veil present, at first covering the gills, then forming a ring on stalk.
6. Flesh very firm, *not* decaying quickly.
7. Growing on ground in late summer or fall (*not* during spring).

Other Features: Cap smooth or cracked; odor mild or cucumberlike but *not* spicy; volva absent; spores white.

Where: On ground under northern and mountain conifers (especially spruce and fir) from the Pacific Northwest to northern California and in the Rocky Mountains and Southwest; common in some areas, otherwise infrequent.

Edibility: Very good, but it needs thorough cooking.

Note: The large "buttons" of this mushroom are among the hardest of all gilled mushrooms. A closely related species, *C. imperialis*, is equally hard but has a browner cap. Both are distinguished from the white matsutake by their decurrent gills and different odor, and from the white alpine waxy cap by their harder flesh and different season. See MD 195 for more information.

White Matsutake
(*Tricholoma magnivelare*)

Other Names: Pine Mushroom, Matsutake, American Matsutake, *Armillaria ponderosa.*

Key Features:
1. Cap white to slightly yellowish (but often with brown or cinnamon stains).
2. Odor strongly spicy-fragrant, like cinnamon or Red Hots (best detected by sniffing the gills).
3. Gills white when fresh (but often with cinnamon stains in age).
4. Gills typically notched (dipping in) where they join the stalk, *not* running down it.
5. Veil present, covering the gills at first, then forming a prominent ring on stalk.
6. Base of stalk *without* a bulb or volva.
7. Growing on ground.

Other Features: Medium-sized to large; flesh very firm; stalk thick, firm, white or with brown or cinnamon-colored scales or fibers below ring; veil rather thick; spores white.

Where: On ground in woods and chaparral, especially in sandy soil under pine, fir, hemlock, Douglas fir, tanoak, madrone, chinquapin, and manzanita. Common from British Columbia to central California; also found in the Rocky Mountains and, rarely, in the Southwest. Harvested for export to Japan.

Edibility: Highly prized for its aroma and complex flavor.

Note: This mushroom, once smelled, is never forgotten. Be careful not to confuse it with white species of *Amanita*, which have a bulb and/or a volva at the base of the stalk, a different odor, and a more fragile ring. See frontispiece and p. 81 for more photos, and MD 191 (as *Armillaria ponderosa*).

Roasted Matsutake

The matsutake's incomparable aroma (above left) is easily lost if it is sauteed like other mushrooms. A far better approach is to slice up the caps and stalks, salt them lightly, and roast or grill the slices (above right) on both sides until very slightly brown. The matsutake will be chewy, but the spicy aroma and complex flavor will permeate your mouth, linger in your throat, and, eventually, go to your head.

Pining for Matsutake

There is a positive correlation between the amount of white matsutake and the size of a matsutake lover's smile (right and below).

Brown Matsutake
(*Tricholoma caligatum*)

Other Names: Fragrant Armillaria, *Armillaria caligata.*

Key Features:
1. Cap covered with brown, reddish-brown, or dark brown fibers or scales.
2. Odor spicy-fragrant (best detected by sniffing the gills), like cinnamon or Red Hots (but see note below).
3. Gills white, typically notched where they join the stalk, *not* running down it.
4. Veil present, covering the gills at first, then forming a prominent ring on stalk.
5. Stalk white above ring, sheathed with brown to dark brown fibers or scales below ring.
6. Base of stalk *without* a bulb or volva.
7. Growing on ground.

Other Features: Medium-sized; veil rather thick; flesh very firm; taste often but not always bitter; spores white.

Where: On ground under northern conifers (especially spruce and fir); fairly common in Alaska, the Rocky Mountains, and the Southwest, but rather rare in the Pacific Northwest and California.

Edibility: Edible but often slightly bitter. When the bitterness is not pronounced it is as good as the white matsutake.

Note: This mushroom resembles the white matsutake except for color. It is also very close to the matsutake of Japan (*T. matsutake*), but the flavor of the Japanese mushroom is superior. One variety of *T. caligatum* with a fruity rather than spicy odor is common in Alaska. See MD 192 (as *Armillaria caligata*) for more information.

Zeller's Tricholoma
(*Tricholoma zelleri*)

Other Names: Fetid Armillaria, *Armillaria zelleri*.

Key Features:
1. Cap orange to yellow-orange or orange-brown, sometimes splashed or mixed with olive-green.
2. Gills white when fresh, attached to stalk but *not* running down it.
3. Veil present, at first covering the gills, then forming a ring on stalk.
4. Stalk with orange or brown stains or scales below the ring.
5. Base of stalk *without* a bulb or volva.
6. Odor unpleasant, like rancid meal or cucumber.
7. Spores white.

Other Features: Medium-sized; cap sticky or slimy when moist; gills often developing orange or brown stains in age; stalk firm, solid, often tapered toward base.

Where: Scattered or in groups on ground in woods; Alaska to central California, and through the Rocky Mountains to the Southwest; especially common under northern and mountain conifers.

Edibility: Not recommended.

Note: The brighter color and unpleasant odor distinguish this mushroom from the brown matsutake and white matsutake, with which it sometimes grows. *T. aurantium* has the same odor and coloration but lacks a ring on the stalk. See MD 187–189 for more information.

Shaggy Stem *(Floccularia straminea)*

Other Names: Scaly Yellow Armillaria, *Armillaria luteovirens, Armillaria straminea.*

Key Features:
1. Cap bright yellow with many yellow protruding or upturned scales (unless washed off by rain).
2. Gills creamy or yellow, notched where they join the stalk or barely attached.
3. Stalk shaggy from a coating of yellow or whitish scales, *without* a bulb or volva at base.
4. Veil present, at first covering the gills, then disappearing or forming a slight ring near top of stalk.
5. Spores white.
6. Growing on ground.

Other Features: Medium-sized; scales on cap sometimes fading to white; stalk *not* snapping open cleanly like chalk.

Where: On ground under aspen or conifers; common in Colorado and adjacent areas, rare or absent elsewhere.

Edibility: Edible; it is popular in Colorado.

Note: The yellow scales on the cap are reminiscent of *Amanita*, but the shaggy stem is distinctive. The edible but mediocre sheathed armillaria (*F. albolanaripes,* below) is very similar but has a smoother, sometimes browner cap. It is common, especially in the mountains, in the spring, summer, and fall. See MD 194 and plate 43.

Yellow Parasol (*Leucocoprinus luteus*)

Other Names: Flower Pot Parasol, *Lepiota lutea, Leucocoprinus birnbaumii.*

Key Features:
1. Cap and stalk bright yellow when fresh, *not* sticky or slimy.
2. Cap distinctly striate (with lines or grooves extending from edge toward center), at least when mature.
3. Cap and stalk covered with fine particles or small scales, at least when young.
4. Gills yellow, close together.
5. Veil present, covering the gills when young, then forming a fragile ring on stalk.
6. Growing indoors, or outdoors only in hot weather.

Other Features: Rather small, withering quickly; cap often fading to tan or whitish as it matures; gills free from stalk; stalk slender; volva absent; spores white.

Where: Common year-round in flower pots and greenhouses, usually in groups or clumps; also in lawns, gardens, and wood chip mulch during hot, humid weather; widespread.

Edibility: Said to be poisonous.

Note: This petite parasol mushroom can be a lovely and informative addition to the household decor when it appears, uninvited, in a flower pot. The photograph was taken in front of a bank in downtown Los Angeles. See MD 302 (as *Lepiota lutea*) for more information.

Green-spored Parasol
(*Chlorophyllum molybdites*)

Other Names: *Lepiota morgani.*

Key Features:
1. Mature cap white with a brown center or with a few scattered brown scales.
2. Gills white or dull greenish, free from stalk.
3. Stalk white (but may have brownish stains), *without* scales, the base often swollen but *without* a large bulb.
4. Veil present, at first covering the gills, then forming a prominent double-edged ring on stalk.
5. Spores greenish.
6. Found primarily on lawns in warm or hot weather.

Other Features: Medium-sized to large; drumstick-shaped with an oval brown or pinkish-brown cap when young, the cap becoming broadly cone-shaped and eventually flat; stalk often long; ring usually movable; volva absent; cut flesh staining orangish, reddish, brownish, or not at all.

Where: On lawns and in other grassy or open places, usually in groups or rings; abundant, especially during the summer, from the Great Plains to the Great Basin, Southwest, southern California, and northern inland California.

Edibility: Poisonous to many people, causing severe gastrointestinal distress (see facing page).

Note: This handsome mushroom is much more likely to grow on lawns than the shaggy parasol or parasol. The greenish spore print is the most reliable feature; the gills may turn greenish or remain deceptively white, even in old age. See MD 295–297 for more information.

The gills of the green-spored parasol can be white, green, or gray-green.

Green Around the Gills

The green-spored parasol is responsible for more poisonings than any other mushroom in North America. In Fresno County, California, 19 cases of mushroom poisoning were reported by local hospitals during one 17-day period, all caused by the green spored parasol! Its prominence on lawns (where those who don't normally eat wild mushrooms are tempted by it) undoubtedly contributes to the plethora of poisonings. Also, it resembles edible mushrooms such as the shaggy parasol and shaggy mane, and is palatable for some people, who recommend it or feed it to others. Fortunately, when gastrointestinal distress ensues it passes within a day or two.

Fairy Rings

Fairy rings occur where there is an even distribution of nutrients in the soil and the mushroom mycelium can spread outward in all directions, growing larger each year and periodically producing mushrooms on its fringes. Many different mushrooms are capable of forming fairy rings, including the green-spored parasol (shown below) and the aptly named fairy ring mushroom (pp. 50–51).

A perfect fairy ring of green-spored parasols.

My Most Memorable Mushroom Hunt

Well, there was the Manteca Coffee Bean Bash, where we gathered hundreds of pounds of oysters in a field full of coffee bean waste — not a tree in sight! And then there was Bitorquis Bounty, when every day for four months we dug up several pounds of prime torks from an area 15 ft. across.

But neither of those can compare with Lepiota Land. Every fungophile has their favorite mushroom, I suppose, but few are blessed with not a patch, not a batch, but a *bonanza* of their favorite. I am sole owner of California license plate LEPIOTA, so forgive me if I wax poetic.

Picture if you will an unsprayed and untended, 15-acre plum orchard, carpeted with massive clumps of *Lepiota rachodes*. Big leppies, small leppies, young leppies, old leppies, but most of all prime, pristine, picture-perfect leppies with not a maggot among them! We just stared in awe at the Lepiota Land of our dreams come true. I thank George for inviting me to Parasol Paradise, and I encourage everyone to keep on trekking — your wildest fungal fantasy may yet come true!

—*Ed Aguilar*

Shaggy Parasol (*Lepiota rachodes*)

Other Names: Leppie, *Lepiota rhacodes*, *Macrolepiota rachodes*.

Key Features:
1. Medium-sized to large: mature cap at least 3" broad; top of stalk usually at least ½" thick.
2. Cap at first smooth and brown but soon breaking up to form shaggy scales on a pale background, the center usually remaining smooth and brown.
3. Flesh staining orange and then reddish when cut (best seen by cutting top of stalk cross-wise).
4. Gills white in all stages, free from the stalk.
5. Veil present, at first covering the gills, then forming a prominent double-edged ring on stalk.
6. Stalk *without* scales; base swollen or with a large bulb.
7. Spores white.

Other Features: Cap at first domed or marshmallow-shaped, flat in old age; ring usually movable, white or brown; stalk white or brownish-stained; bulb sometimes volva-like.

Where: In groups or clusters (often in large numbers) on ground in woods, orchards, old pastures, gardens, under cypress, near roads, stables, anthills, piles of grass clippings, etc.; widespread and common (year-round in some areas).

Edibility: Excellent, but some people are adversely affected by it. To help reduce the chances of gastrointestinal upset, saute it on high heat in an open pan.

Note: This handsome mushroom can be confused with the poisonous green-spored parasol, but has white spores and favors cooler weather. Old, brown-fleshed specimens may not show the distinctive staining reaction. See pp. *xx* and 81 for more photos, and see MD 297–298.

The parasol mushroom, *Lepiota procera*, is one of our tallest and most delicious mushrooms.

Parasol (*Lepiota procera*)

Other Names: *Macrolepiota procera.*

Key Features:
1. Medium-sized to large: mature cap at least 3" broad.
2. Cap at first smooth and brown but soon breaking up to form shaggy scales on a pale (white, tan, or grayish) background, the center usually remaining smooth and brown.
3. Flesh *not* staining orange or red when cut and rubbed.
4. Gills white in all stages, free from stalk.
5. Veil present, at first covering the gills, then forming a prominent double-edged ring on stalk.
6. Stalk tall and slender (5–20" long, usually ⅜–⅝" thick at the top), the base *without* a volva.
7. Stalk surface covered or belted with delicate brown scales (but these sometimes wearing away in age).
8. Spores white.

Other Features: Drumstick-shaped (with an oval cap) when young, the cap becoming flat or nearly so at maturity; stalk usually thicker at base but *without* a large bulb; ring usually movable

Where: Alone or in groups on ground in open woods and old pastures, along roads, etc. Common throughout most of the northern hemisphere, but in western North America presently known only from southern Arizona.

Edibility: One of the best! The large, thin caps can be fried or broiled whole, or sliced up.

Note: This handsome mushroom is much taller than the shaggy parasol and does not stain orange or red when cut. Its white spores and delicately scaly stalk distinguish it from the green-spored parasol. See MD 298–299 for more information.

Ma'am On Motorcycle
(*Lepiota naucina*)

Other Names: Woman on Motorcycle, Smooth Parasol, Smooth Lepiota, *Leucoagaricus naucinus*, *Lepiota naucinoides*.

Key Features:
1. Cap white to pale gray, bald, *not* sticky or slimy.
2. Gills white or slightly pinkish, free from stalk.
3. Stalk white (sometimes with brownish stains), *without* scales, thickened or swollen at base.
4. Veil present, covering the gills at first, then forming a ring on stalk.
5. Volva absent (no sack, cup, or collar at base of stalk).
6. Spores white.
7. Growing in grass.

Other Features: Medium-sized; cap egg-shaped or helmet-shaped when young, becoming domed and eventually flat; cap and stalk usually staining brown when handled or cooked (but one variant staining yellow); gills close together.

Where: In lawns, pastures, and other open places, often in groups; widespread and common.

Edibility: Edible, but some people are made ill by it. Also, be careful not to confuse it with a deadly amanita!

Note: This common, stately suburbanite lacks the brown scales of most parasol mushrooms. There is never a sack or volva at the base of the stalk as in the destroying angel (*Amanita ocreata*). The young caps are reminiscent of motorcycle helmets, and are easily confused with *Agaricus*. See MD 299–300 for more information.

Shaggy-stalked Parasol
(*Lepiota clypeolaria*)

Other Names: *Lepiota ventriosospora.*

Key Features:
1. Cap usually less than 3" broad when expanded.
2. Mature cap with a smooth brown center and scattered small brown scales on a pale background, the edge usually ragged from hanging veil remnants.
3. Stalk slender (usually <⅜" thick), shaggy or woolly, fragile.
4. Gills white or creamy, free from stalk.
5. Cottony veil present, covering gills when young but forming only a slight ring on stalk (if any) after breaking.
6. No part of mushroom staining when bruised or cut.
7. Base of stalk with neither bulb nor volva.
8. Growing in woods.

Other Features: Cap oval or rounded at first, becoming flat at maturity; surface at first brown or yellow-brown but soon breaking up to form scales on a white or yellowish background; stalk colored like cap or yellower; spores white.

Where: On ground in woods, often in groups; widespread, but especially common on the West Coast.

Edibility: Poisonous.

Note: The cap, which is often suggestive of a human breast, and the shaggy stalk are distinctive. This and other small species of *Lepiota* should not be eaten, as some are dangerously poisonous and accurate identification is difficult. See MD 304–310 for more information.

Agaricus albolutescens
(see "Note" on facing
page).

6 Agaricus

This important group, which includes the familiar supermarket variety, is easily recognized by the features shown in the above photo: the gills are free from the stalk and dark chocolate-brown at maturity (but white or pink when young), and they are covered at first by a veil which usually forms a ring on the stalk after breaking. In addition, the spores are the color of unsweetened chocolate (the ring in the photo is coated with spores).

Distinguishing among different species of *Agaricus* can be tricky, because most have the same overall appearance (white or brown cap, white stalk, and the above-mentioned features). As both delectable and mildly poisonous species (see p. 109) occur, a large selection is included here. It is easily the most difficult section of this book, requiring a keen eye and sensitive nose, so seek out experienced help if possible. Your efforts will be rewarded, because no group of gilled mushrooms offers more fine edibles, many of which are plentiful in urban as well as rural habitats. For a more comprehensive treatment, see MD 310–341.

Woodland Agaricus
(*Agaricus silvicola*)

Key Features:
1. Cap white to yellowish, *not* sticky or slimy.
2. Edge of cap staining *and remaining* yellow when surface is rubbed repeatedly.
3. Gills free from stalk, pale when young, eventually becoming dark chocolate-brown.
4. Odor of crushed flesh sweet (like almond extract or anise), especially when young.
5. Stalk usually partly hollow or stuffed with a pith.
6. Veil present, at first covering the gills, then forming a skirtlike ring on stalk.
7. Base of stalk *without* a volva (cup or sack), the tip yellowing only slightly if at all when cut.
8. Spores chocolate-brown.
9. Growing in woods.

Other Features: Medium-sized; cap usually bald; stalk white, smooth or slightly scaly, with or without a bulb at base; veil often with patches on underside.

Where: Alone or in small groups on ground in woods and at their edges; common and widespread.

Edibility: Delicious, but be sure not to confuse it with the deadly poisonous destroying angel!

Note: Several edible woodland species have the above features. One, *A. albolutescens* (shown at top of facing page), is very fragrant and quickly stains amber when bruised. The poisonous yellow stainer has a phenol odor and yellow-staining stalk base, while the horse mushroom grows in grass. See MD 334–336 for more information.

Giant Horse Mushroom
(*Agaricus osecanus* & *A. nivescens*)

Key Features:
1. Medium-sized to large (cap >3" broad, stalk >¾" thick).
2. Cap white to slightly yellowish, *not* sticky or slimy.
3. Edge of cap yellowing only slightly if at all when surface is rubbed repeatedly.
4. Gills free from stalk, pale when young, eventually becoming dark chocolate-brown.
5. Veil present, at first covering the gills, then forming a prominent ring on stalk.
6. Stalk solid or nearly solid inside.
7. Base of stalk *without* a volva (cup or sack), the tip *not* staining bright yellow when cut.
8. Growing in grass.

Other Features: Cap smooth or warty; odor of crushed flesh faintly sweet when young; stalk bald or with a few scales, *without* a bulb at base (but may be swollen above base); spores chocolate-brown.

Where: In pastures and other grassy or open places, often in groups; common on West Coast (especially during spring in coastal pastures), perhaps more widely distributed.

Edibility: Delicious, especially when young and firm.

Note: The above features apply to several large edible mushrooms of open places that differ microscopically (including *A. crocodilinus*, shown at bottom of facing page). As a rule they are more robust than the horse mushroom (p. 104) but not as fragrant. The cap can be smooth or warty depending on exposure to wind and sun; the tip of the stalk never stains bright yellow as in the yellow stainer. See MD 333–334 for more information.

Have Fence, Will Trespass

I was speeding up the coast, late for an important date. The lush green hills were brushed dark to light by gusts of wind like a pool table before the opening break. It was while scanning the meadows, thinking of Ireland, that I saw them: salt licks? paper plates? a flock of migrating geoducks? Some people use binoculars for confirmation, others prefer pellet guns and see if the sightings go baaaaaa . . . but I just *knew*. So what if I stopped? I was *already* late, and they were beauties: white, naked, shivering in the wind. The only thing between them and my belly was a mean little barbed wire fence , , ,

—*El Bo*

The crocodile agaricus, *Agaricus crocodilinus* (see "Note" on facing page).

Horse Mushroom (*Agaricus arvensis*)

Key Features:
1. Cap white to yellowish, medium-sized (mostly 2–6" broad), *not* sticky or slimy.
2. Edge of cap staining *and remaining* yellow when surface is rubbed repeatedly.
3. Gills free from stalk, pale when young, eventually becoming dark chocolate-brown.
4. Odor of crushed flesh sweet (like almond extract or anise), especially when young.
5. Stalk usually partly hollow or stuffed with a pith.
6. Veil present, at first covering the gills, then forming a skirtlike ring on stalk.
7. Base of stalk *without* a volva (cup or sack), the tip *not* staining bright yellow when cut.
8. Growing in grass.

Other Features: Cap usually bald; stalk white, bald or with a few small scales; underside of unbroken veil usually with a cogwheel pattern of patches; spores chocolate-brown.

Where: Common in lawns, cemeteries, golf courses, and other grassy areas, less frequent in pastures; widespread.

Edibility: Delicious; it is sweeter than the giant horse mushroom, but don't confuse it with a poisonous amanita!

Note: This popular edible species is usually smaller, yellower, and more fragrant than the giant horse mushroom. It resembles the woodland agaricus but grows in grass. On lawns it often mingles with poisonous species of *Agaricus*, but the sweet odor and permanent yellowing of the cap distinguish it. See MD 332–333 for more information.

Almond Mushroom
(*Agaricus subrufescens*)

Key Features:
1. Cap white to tan with small light brown fibers or scales.
2. Edge of cap staining *and remaining* yellow when surface is rubbed repeatedly.
3. Odor strongly sweet (like almond extract), at least when flesh is crushed.
4. Gills free from stalk, pale when young, dark chocolate-brown when older.
5. Stalk *not* sheathed with prominent scales.
6. Veil present, at first covering the gills, then forming a skirtlike ring on stalk.
7. Base of stalk *without* a volva, staining pale yellow or orange-yellow when crushed.
8. Growing in grass or rich soil but *not* in woods.

Other Features. Medium-sized to fairly large; unbroken veil with patches on underside; spores chocolate-brown.

Where: In lawns, gardens, and manured soil during warm weather; common in southern California, otherwise rare.

Edibility: Excellent — the most almondy agaricus, even sweeter than the prince!

Note: This mushroom can be grown at home in the shade of leafy summer vegetables. It was cultivated commercially at the turn of the century but lost out to the "supermarket mushroom," *A. bisporus.* The robust, clustered specimens in the photograph grew in manure; those in grass are usually smaller and scattered. See MD 336–337 for more information.

Above: "Spore prince": a fully mature *A. augustus* shedding its spores. **Right**: This woman has found her Prince Charming.

A Regal Repast

"August: Inspiring Awe or Admiration"

This majestic mushroom is my favorite, a culinary and aesthetic delight. The flavor is strong, nutty, and sweet like almonds, and if you are blessed with a cluster, it will make the best cream of mushroom soup on the planet.

Wipe the mushrooms clean, slice thinly (don't soak in water!), and saute in butter or a mild oil. Pour off and save the liquid essence they produce, and continue to cook until almost dry. Douse with David Bruce Santa Cruz Mountains Chardonnay (just a splash; save the rest to sip while you eat) and cook again until *almost* dry. Sprinkle with a little flour. Cook 1 minute, stirring constantly, and add the reserved essence plus a bit more water or *mild* stock. Simmer ½ hour, uncovered, allowing it to reduce. Add salt, white pepper, and milk or cream to taste (excellent also without the milk or cream), then serve or reduce further for a delicious sauce over pasta, fish, or vegetables. And please invite me over for dinner!

—*Judith Scott Mattoon*

A princely cluster; note the shaggy stalks.

Prince (*Agaricus augustus*)

Key Features:
1. Cap with brown to golden-brown fibers or scales, at least 3" broad when mature.
2. Cap marshmallow-shaped when young (before veil breaks).
3. Edge of cap staining *and remaining* yellow when surface is rubbed repeatedly.
4. Odor sweet (like almond extract), at least when flesh is crushed.
5. Gills free from stalk, white when young, dark chocolate-brown in old age.
6. Veil present, white, covering the gills at first, then forming a skirtlike ring on stalk.
7. Stalk at least ½" thick, sheathed with scales below the veil (at least when young).
8. Base of stalk *without* a volva, the tip yellowing only slightly if at all when cut.

Other Features: Medium-sized to very large; cap eventually flat, *not* sticky or slimy; base often deeply buried; veil with patches on underside; spores chocolate-brown.

Where: On ground in woods (usually near roads or trails) and in disturbed areas or rich soil, often in groups or clumps; widespread but rare except on the West Coast. In the coastal fog belt it fruits in the summer when other fungi are scarce.

Edibility: One of the very best! The sweet fragrance and flavor are a real treat.

Note: The almondy odor, shaggy stalk (see photo at bottom of facing page), and yellow staining of the cap are most obvious when young. Frisbee-sized mature caps are not uncommon. *A. smithii* and *A. perobscurus* are closely related edible species. See MD 337–340 for more information.

107

Flat-topped Agaricus
(*Agaricus praeclaresquamosus*)

Other Names: *Agaricus meleagris.*

Key Features:
1. Cap with brown to inky gray fibers or small scales (at least at the center), *not* sticky or slimy.
2. Young cap marshmallow-shaped (with flattened top).
3. Gills free from stalk, whitish when young (sometimes pink *after* veil breaks), then dark chocolate-brown.
4. Stalk *without* scales, often curved above base.
5. Veil present, at first covering the gills, then forming a prominent ring on stalk.
6. Base of stalk *without* a volva, the tip typically staining *bright* yellow when cut (see photo on facing page).
7. Crushed flesh with an unpleasant antiseptic odor (like phenol or paste), especially in base of stalk.
8. Growing in woods or near trees.

Other Features: Medium-sized to large; cap eventually flat, surface occasionally yellowing when rubbed (but then turning brownish later); veil fairly thick; spores chocolate-brown.

Where: On ground in woods and at their edges, and under trees, often in groups; widespread, but especially common in California and the Pacific Northwest.

Edibility: Poisonous to many, causing stomach distress.

Note: This handsome mushroom is sometimes mistaken for the prince. It does not smell almondy, however, and is smooth-stalked rather than shaggy. One variant does not stain bright yellow in the stalk base, but can still be recognized by the other features listed above. For more details, see facing page and MD 329.

A. praeclaresquamosus (facing page)

A. hondensis (p. 114)

A. xanthodermus (p. 110)

A. californicus (p. 111)

The Lose Your Lunch Bunch

These four poisonous species of *Agaricus* are frequently mistaken for edible look-alikes. Also known as the "Foolgoof Four" (as opposed to the so-called "Foolproof Four"; actually, no mushrooms are foolproof, though dozens are intelligence-proof), they can be distinguished from edible species of *Agaricus* by their unpleasant odor (like phenol or library paste, strongest when the base of the stalk is crushed). Also, the stalk base often has a crook above it, and in two of the four it stains bright yellow when nicked.

Some people with a poor sense of smell prefer to use a simple chemical test: a weak solution of Drano or potassium hydroxide stains the cap surface of the Lose Your Lunch Bunch yellow, but does *not* stain the cap of the meadow mushroom (*A. campestris*) and several close relatives. This same test, however, also produces a yellow reaction in the fragrant edible species that naturally stain yellow (such as the prince and horse mushroom). Fortunately, should you munch on one of the Lose Your Lunch Bunch, the consequences aren't serious: stomach distress that subsides once the ingested offender has been expelled.

Yellow Stainer (*Agaricus xanthodermus*)

Other Names: Yellow-staining Agaricus, Yellow-foot.

Key Features:
1. Cap white to tan, *not* sticky or slimy.
2. Edge of cap staining bright yellow when surface is rubbed repeatedly, then later staining brownish.
3. Gills free from stalk, white at first, becoming pinkish *after* veil breaks, then dark chocolate-brown.
4. Stalk *without* scales, often curved above base.
5. Veil present, at first covering the gills, then forming a prominent ring on stalk.
6. Base of stalk *without* a volva, the tip instantly staining *bright* yellow when cut.
7. Crushed flesh with an unpleasant antiseptic odor (like phenol or paste), especially in base of stalk.

Other Features: Medium-sized; young cap domed or marshmallow-shaped, the edge usually tucked under when young; cap surface bald or somewhat scaly; spores chocolate-brown.

Where: Widespread in lawns, gardens, woods, orchards, pastures, and along roads, often in large groups; one of the most abundant urban and rural mushrooms in California.

Edibility: Poisonous to many, causing gastrointestinal distress. The cooked mushroom smells and tastes foul.

Note: Several species of *Agaricus* stain yellow, but none more dramatically than this one. The extreme base of the stalk always stains bright yellow when nicked, and the cap surface typically yellows also. See p. 109 for another photo and more details; also see MD 329–331.

Mock Meadow Mushroom
(*Agaricus californicus*)

Other Names: California Agaricus, Fool's Agaric.

Key Features:
1. Cap whitish to brown, bald or with fibers or scales, *not* sticky or slimy.
2. Edge of cap *not* yellowing when surface is rubbed repeatedly, usually tucked under when young.
3. Gills free from stalk, at first whitish, becoming pink *after* veil breaks, then dark chocolate-brown.
4. Flesh staining slightly if at all when cut and rubbed.
5. Stalk *without* scales, usually <½" thick at top, often curved above base.
6. Veil present, at first covering gills, then forming a prominent, persistent ring on stalk.
7. Base of stalk often swollen but *without* a volva, the tip *not* staining bright yellow when cut.
8. Base of stalk with slightly unpleasant antiseptic odor when crushed (like phenol or paste).

Other Features: Small to medium-sized; young cap domed or marshmallow-shaped; cap often whitish with a brown center, often with metallic sheen; spores chocolate-brown.

Where: In lawns, gardens, under trees, or in woods; abundant in California but probably more widespread.

Edibility: Poisonous to many, causing stomach upsets.

Note: This ubiquitous species can be mistaken for the meadow or brown meadow mushroom but has a prominent ring, frequently crooked stalk, whitish gills *before* the veil breaks, and slightly unpleasant odor. Other members of the "Lose Your Lunch Bunch" are larger. See p. 109 and MD 327–328.

Brown Meadow Mushroom
(*Agaricus cupreobrunneus*)

Other Names: Brown Field Mushroom.

Key Features:
1. Cap brown and fuzzy or hairy, *not* sticky or slimy.
2. Gills free from stalk, pink *before* and soon after veil breaks, then becoming dark chocolate-brown.
3. Veil present, at first covering gills, but forming little or no ring on stalk; ring, if present, *not* skirtlike.
4. Stalk typically *without* a crook above base; base *without* a volva (cup or sack) or bulb.
5. Base of stalk with a mild or mushroomy odor when crushed (i.e., *not* noticeably unpleasant or sweet).
6. Extreme base (tip) of stalk *not* staining yellow when cut.
7. Growing in grass.

Other Features: Small to medium-sized; cap domed at first, *not* yellowing when rubbed, edge projecting beyond gills; flesh fragile, *not* typically staining when cut but may redden slightly with age or wet weather; spores chocolate-brown.

Where: In pastures and other grassy or open places, often in large numbers; widespread, but especially abundant on West Coast. It often grows with the meadow mushroom but seems better suited to poor soils.

Edibility: Edible but fragile and perishable.

Note: Though plentiful, this mushroom is smaller, more fragile, and often dirtier than the meadow mushroom. See MD 319 for more information.

> An *Agaricus* fancier named Clyde
> (Loved *campestris* baked, raw, even deep-fried)
>   Said, "My, I really feel rotten —
>   I fear that I've forgotten
> To test with KOH!"
>
> —Charles Sutton

Meadow Mushroom
(*Agaricus campestris*)

Other Names: Field Mushroom, Pink Bottom, Champignon.

Key Features:

1. Cap white, sometimes with scattered brown fibers or scales, *not* sticky or slimy.
2. Gills free from stalk, pink *before* and soon after veil breaks, then becoming dark chocolate-brown.
3. Flesh *not* staining appreciably when cut and rubbed.
4. Veil present, at first covering gills, but forming little or no ring on stalk; ring, if present, *not* skirtlike.
5. Stalk typically *without* a crook above base.
6. Base of stalk *without* a volva (sack or cup) or bulb, the tip *not* staining yellow when cut.
7. Base of stalk with a mild or mushroomy odor when crushed (i.e., *not* noticeably unpleasant or sweet).
8. Growing in grass.

Other Features: Medium-sized; young cap domed; cap not normally yellowing when rubbed, edge projecting beyond gills; stalk usually tapered at base; spores chocolate-brown.

Where: In lawns, cemeteries, alpine meadows, and other grassy areas, often in groups or rings; cosmopolitan. "Champignonship" crops most often appear in grazed pastures.

Edibility: A popular favorite, even among staunch fungophobes such as the English.

Note: The pink-gilled "buttons" are as lovely as they are delicious. Unfortunately, on California lawns they often mingle with poisonous look-alikes. In the mock meadow mushroom and yellow stainer there is a ring on the stalk and often a crook near the base, and the gills do not turn pink until *after* the veil breaks. See MD 318 for more information.

Felt-ringed Agaricus
(*Agaricus hondensis*)

Key Features:
1. Cap with pale tan, brown, or reddish-brown fibers or small flattened scales; surface *not* sticky or slimy and edge *not* yellowing quickly when rubbed.
2. Gills free from stalk, whitish or pinkish when young, dark chocolate-brown when older.
3. Flesh *not* reddening immediately when cut and rubbed.
4. Stalk *without* scales, at least ½" thick at top, often with a crook above base.
5. Veil present, at first covering the gills, then forming a thick, felty ring on stalk.
6. Base of stalk *without* a volva, the tip staining yellow only slightly if at all when cut.
7. Base of stalk with an unpleasant antiseptic odor when crushed (like phenol or paste).
8. Growing in woods or under trees.

Other Features: Medium-sized; cap domed at first; stalk base usually thickened or swollen; spores chocolate-brown.

Where: On ground in woods and under trees, often in groups or rings; common on the West Coast. It tolerates cold weather better than most species of *Agaricus*.

Edibility: Poisonous to many people, causing gastrointestinal distress. It develops a strong phenol odor while cooking and has an unpleasant, slightly metallic taste.

Note: This handsome forest mushroom has several look-alikes with reddening flesh and a pleasant odor. The wine-colored agaricus is also similar but has a shaggy stalk. For another photo and more details, see the "Lose Your Lunch Bunch" on p. 109, and MD 326–327.

Wine-colored Agaricus
(*Agaricus subrutilescens*)

Other Names: Woolly-stemmed Agaricus.

Key Features:
1. Cap covered with purple-brown or wine-brown fibers or scales, *not* sticky or slimy, *not* yellowing when rubbed.
2. Gills free from stalk, whitish at first, then becoming pink and finally dark chocolate-brown.
3. Flesh *not* reddening or yellowing when cut and rubbed repeatedly.
4. Veil present, at first covering the gills, then forming a thin skirtlike ring on stalk.
5. Stalk sheathed with shaggy white scales below veil.
6. Base of stalk *without* a volva, the tip *not* staining bright yellow when cut.
7. Odor mild or pleasant when base of stalk is crushed (*not* smelling like phenol, paste, or almond extract).
8. Growing in woods.

Other Features: Medium-sized; stalk thick or slender; spores chocolate-brown.

Where: On ground in woods and at their edges; common from the rain forests of the Pacific Northwest to the oak and pine woodlands of southern California.

Edibility: Edible for most people but mildly poisonous to some (causing gastrointestinal distress).

Note: Both slender and robust forms occur. The distinctive shaggy white sheath on the stem is most obvious when young, before the mushroom is handled. In older specimens the sheath may be obliterated, leading to confusion with the poisonous felt-ringed agaricus or other species. See MD 326 and plate 75.

Alpine Bleeder (*Agaricus amicosus*)

Key Features:

1. Medium-sized (cap 2–7" broad, top of stalk usually <1" thick).
2. Cap covered with tan to brown fibers or scales, *not* sticky or slimy.
3. Gills free from stalk, pinkish when young, eventually becoming dark chocolate-brown.
4. Flesh reddening when cut and rubbed repeatedly (especially at juncture of cap and stalk).
5. Veil present, covering the gills at first, then forming a skirtlike ring on stalk.
6. Base of stalk *without* a volva, the tip *not* staining bright yellow when cut.
7. Growing under mountain conifers.

Other Features: Stalk bald or slightly scaly, hollow or stuffed with a pith; odor of crushed flesh mild or pleasant (*not* like phenol or paste); spores chocolate-brown.

Where: On ground under spruce and fir, often in groups or rings; presently known from the mountains of Colorado and New Mexico, where it is common, but probably found in adjacent states.

Edibility: Presumably edible.

Note: There are several medium-sized, red-staining *Agaricus* species (or "bleeders") with a skirtlike ring. They can be difficult to differentiate, but fortunately none are known to be poisonous. Three are shown on the facing page; others include *A. fuscofibrillosus* (MD plate 74), common in coastal California under cypress, and *A. silvaticus*, found under conifers in the Pacific Northwest. See MD 324–326 for more information.

A. fuscovelatus is fairly common in California, especially under cypress. It can be recognized by its purplish or chocolate-colored veil and tendency to stain orangish or reddish (at least slightly) when cut.

Other Bleeders

A. arorae is a small, slender inhabitant of coastal California that "bleeds" readily like its namesake when cut. It is closely related to *A. amicosus* in that both stain yellow in potassium hydroxide, whereas most "bleeders" do not.

This firm, robust, delicious mushroom is common along roads in Alaska. It is *Agaricus vaporarius*, or a similar, perhaps unnamed species.

Giant Cypress Agaricus
(*Agaricus lilaceps*)

Key Features:
1. Medium-sized to large (mature cap at least 3" broad).
2. Cap normally brown (but see note below), bald or with fine fibers, *not* sticky or slimy.
3. Gills free from stalk when mature, pinkish when young, dark chocolate-brown when older.
4. Stalk at least 1" thick, of equal width throughout or slightly thicker below, *not* scaly.
5. Flesh thick, firm, staining at least slightly reddish or dull burgundy when cut and rubbed repeatedly.
6. Veil present, covering gills for a long time (until mature), finally forming a thin skirtlike ring on stalk.
7. Base of stalk *without* a bulb or volva, *not* staining bright yellow when cut.
8. Base of stalk with mild or fragrant odor when crushed (*not* unpleasant).
9. Usually growing near cypress.

Other Features: Young cap broadly domed; unbroken veil often with yellow or brown patches; odor of crushed flesh mild or fragrant (*not* unpleasant); spores chocolate-brown.

Where: On ground under mature Monterey cypress or sometimes other trees, often half-buried, uncommon. At present it is known only from coastal California, but perhaps readers will find it in new localities!

Edibility: Delicious — one of the best!

Note: This marvelous, meaty mushroom deserves to be better known. Under certain conditions the normally brown cap can develop striking orange, yellow, pink, or lilac hues. Another robust red-stainer, *A. pattersonae*, has a scalier stalk with a bulb. *A. bernardii* can be large, but has a different kind of ring and habitat; the poisonous *A. hondensis* often grows under cypress but has a thicker ring and unpleasant odor. See MD 323–324 for more information.

Salt-loving Agaricus
(*Agaricus bernardii*)

Key Features:
1. Cap white to tan, *not* sticky or slimy.
2. Gills free from stalk at maturity, pinkish or brown *before* veil breaks, dark chocolate-brown when older.
3. Flesh thick, very firm, staining orange, reddish, or dull wine-red when cut and rubbed repeatedly.
4. Stalk hard, solid (neither hollow nor stuffed with a pith).
5. Veil present, white while covering the gills, forming a ring on stalk.
6. Ring *not* hanging down like a skirt.
7. Base of stalk *without* a volva (sack or cup) or bulb, *not* staining bright yellow when cut.

Other Features: Medium-sized to fairly large; cap bald, scaly, or warty, the edge tucked under when young and projecting beyond gills when mature; odor of crushed flesh mild, mushroomy, or briny; spores chocolate-brown.

Where: On lawns (often in rings) and in hard-packed or disturbed soil; widespread, but especially common in California. It hugs the ground closely or develops beneath it.

Edibility: Underrated; it sometimes has a briny taste but is usually excellent.

Note: This mushroom is closely related to *A. bitorquis*, but reddens when cut and often has a scalier cap. A similar edible reddening species with brown fibers or scales on the cap (*A. vaporarius?*) is shown at the bottom of p. 117. Other red-stainers have a skirtlike ring and different habitat. See MD 322 for more information.

Above: Ed Aguilar harvested 76 torks from this patch, but only a few are visible in this photo. That's because the rest were underground, and had to be dug up like clams (see pp. 122–123). Torks don't normally grow in pastures, but this one was covered with a layer of silt from a recent flood.

Left: Knife and tork.

"We ate fabulously and got along famously" (see "Mining for Mushrooms," pp. 122–123). Torks are superb stuffed and broiled or sauteed and served over pasta.

Tork (*Agaricus bitorquis*)

Other Names: Banded Agaricus, Urban Agaricus, Spring Agaricus, *Agaricus rodmani*.

Key Features:
1. Cap white to pale tan, bald, *not* yellowing when rubbed repeatedly, *not* sticky or slimy.
2. Gills free or nearly free from stalk, pinkish *before* veil breaks, dark chocolate-brown in age.
3. Flesh firm, staining only slightly if at all when cut.
4. Stalk hard, solid (neither hollow nor stuffed with a pith).
5. Veil present, at first covering the gills, then forming a prominent ring on stalk.
6. Ring bandlike (with both edges free or flaring) or sheathing (with only upper edge flaring), *not* skirtlike.
7. Base of stalk with a mild odor (*not* sweet or unpleasant) when crushed, *not* staining yellow when cut.
8. Developing underground (but often surfacing later).

Other Features: Medium-sized to fairly large; edge of cap tucked under when young, projecting beyond gills when older; flesh thick; stalk *without* a bulb at base; ring sometimes resembling a volva; spores chocolate-brown.

Where: Usually buried or half-buried in hard-packed or disturbed soil, silt, mud, or sometimes grass; widespread. In regions with cold winters it is most plentiful in the spring.

Edibility: Very firm and flavorful. One devotee goes so far as to say: "Life is like an *Agaricus bitorquis* — all good except for a few gills." Life may not be so consistently good, but *A. bitorquis* certainly is!

Note: The names spring and urban agaricus reflect the preferred season and milieu, but several kinds of *Agaricus* are urban or grow in the spring. The bandlike or sheathing ring is more distinctive. See MD 321–322 for more information.

Mining for Mushrooms

(to be said aloud)

It was a long time ago, in my hippie days. I was living on a commune, and I was sick and tired of all the bickering and brown rice. I really needed some *space*, so I split for Arizona, where I heard there was nothing *but*, to see the spring wildflowers. So get this: we're driving down this crusty, dusty desert road on the way to a scenic overlook — *the most unlikely place in the world for mushrooms* — and I see this glimmer of white in the ditch by the road. We stop for a look and, sure enough, it's an old *Agaricus bitorquis*. Jade says it must be the only shroom in the state of Arizona, and I'm about to agree when I start noticing all these *cracks* everywhere in the hard red clay along the road. It was *shroom city*. There were *hundreds*, big clumps of them, *veins* of them, but all underground! Most were several inches under, some more than a foot. "Dig this!" I said to Jade. "With what?" she wanted to know. We used our hands, making piles of them on the road as we walked along.

Of course we were noticed. An RV stopped, and this older couple from Long Beach got out and wanted to know what we were doing. "*We're mining for mushrooms,*" I said, pausing for effect, "*and we've just struck the mother lode.*" We could tell they really wanted to try their hand at it. They sold life insurance and had been traveling for three months, visiting *every* national park in the country and this was their *final* stop, their *last* scenic overlook, and they were so burned out, they really wanted to do something *exciting*. But duty called, they just *had* to go on to the overlook.

Five minutes later they were back for some *fun*. Along with everything else in the world they had brand new shovels with them which they'd been wanting to use for months,

"It was the most unlikely place in the world for mushrooms . . ."

"Then I started noticing all these *cracks* everywhere in the hard red clay."

and they started pulling giant buttons out of the ground like clams. Boy were they stoked! Mushrooms, *edible* mushrooms, under the sun-baked desert crust! It was totally incredible to them. It wasn't in their tourist guides or on their itinerary, the auto club hadn't said anything about it, it had never occurred to them to eat wild mushrooms, so they just got more and more excited and started scurrying around yelping and babbling like kids, "Look at this sonofagun over here!"; "Mine's even bigger than yours!"; "Holy Cow, it's hard as a rock!"; "I can't believe I'm doing this!"

Another RV pulled over to see what all the commotion was about. One of them also sold insurance and of course they had shovels, so they dug right in. Then another RV joined us, a Mormon family from Moab, a bicyclist bound for Lubbock, and two local Navajo. We must have pulled up a couple hundred pounds, and we left *lodes* behind. Talk about "overlook" — we wouldn't have gotten *any* if that one old cap hadn't made it above the ground!

There was only one campground in the area and we were all staying there, so that night we had this incredible spontaneous mushroom feast with gourmet foods and drinks they'd stashed away in their RVs for that one really special occasion, and what could be more special than this? We ate fabulously and got along famously, and the couple from Long Beach wanted to know if this was what it was like to live communally and I said: "Sure, we do this every night."

I guess you could say we made their day. In fact, they said it was the best thing that happened on their whole trip! We had more for breakfast the next morning, and sun-dried the rest, and that one couple just couldn't stop talking about how excited they were. I kept getting letters from them afterwards, and I bet they're *still* talking about it, twelve years later, telling their grandchildren about the mighty once-in-a-blue-moon shroom bloom beneath the Arizona desert. Me, I'm not much of a talker, but I'm sure tempted to go back — I never did make it to that scenic overlook!

—Max Lipp

Above: The big and small of it: each *Coprinus disseminatus* (see MD 352) averages ¼" broad; king stropharias (facing page) can weigh several pounds. **Below:** Gypsies (see p. 140, and story on p. 189).

7 Miscellaneous Dark-spored Gilled Mushrooms

All gilled mushrooms with deeply colored spores are grouped here with the exception of *Agaricus*. The spore color ranges from rusty-orange to brown, rusty-brown, purple-gray, or black. A veil may or may not be present. In those with a veil, either the gills are attached to the stalk or the spores are differently colored than those of *Agaricus*.

There are more than 1000 western mushrooms in this category. A few are sought for their edible or hallucinogenic properties, but most are poisonous or of unknown edibility. See MD 341–487 for a more comprehensive treatment.

King Stropharia
(*Stropharia rugoso-annulata*)

Other Names: Wine-red Stropharia, Giant Garden Stropharia.

Key Features:
1. Cap wine-red to red-brown when young (but often fading to tan as it ages).
2. Surface of cap sticky or slimy when moist, bald.
3. Gills attached to stalk, at first gray or purple-gray, becoming black in old age.
4. Stalk at least ½" thick, white when fresh.
5. Veil present, at first covering the gills, then forming a prominent ring on stalk.
6. Ring grooved or ridged on upper surface.
7. Spores deeply colored (dark purple-brown to black).

Other Features: Medium-sized to very large; flesh white, *not* discoloring appreciably when bruised; ring often segmented or clawlike; white threads often emanating from base of stalk; volva absent.

Where: In humus, wood chips, straw, grass, gardens, landscaped areas, etc., often in groups; fairly common in the Pacific Northwest. As it is grown at home by more and more people, its range will surely spread.

Edibility: Very good: cook it as you would the regular commercial mushroom.

Note: This handsome mushroom has become very popular with home growers because it is easy to cultivate, attains prodigious size (see photo on facing page), fruits prolifically, and is tasty besides! See MD 378–379 for more information.

Questionable Stropharia
(*Stropharia ambigua*)

Key Features:
1. Cap buff to yellow or yellow-brown.
2. Surface of cap sticky or slimy when moist, bald.
3. Edge of cap adorned with cottony white veil tissue.
4. Mature gills gray to purple-gray or black, usually attached to stalk.
5. Stalk with a shaggy coating of soft white scales.
6. Veil present, at first covering gills, then shredding into many pieces that dangle from edge of cap, and sometimes also forming a slight ring on stalk.
7. Spores deeply colored (dark purple-brown to black).

Other Features: Medium-sized, usually tall and slender; cap domed at first, flat in age; gills occasionally pulling away from stalk in age; white threads often emanating from base; volva absent.

Where: On ground, rich soil, or wood chips in forests and at their edges; common on West Coast. It favors cold, damp situations.

Edibility: Not recommended.

Note: This is an elegant mushroom in its prime. The bald, sticky or slimy cap and attached gills distinguish it from *Agaricus*, while *Amanita* has much paler gills. *S. hornemanni* is a similar woodland species with a yellow-brown, red-brown, or gray-purple cap and well-developed ring on the stalk. See MD 377–380 and plate 89.

Dung Dome (*Stropharia semiglobata*)

Other Names: Hemispherical Stropharia, Round Stropharia.

Key Features:
1. Cap small (<2" broad and usually ±1"), domed when young and slightly domed even when mature.
2. Cap yellow, blond, or straw-colored, the surface bald and sticky or slimy when moist.
3. Gills gray to black when mature, attached to stalk.
4. Stalk slender, *without* scales, at least the lower half sticky or slimy when moist.
5. Veil present in young specimens.
6. Cap, stalk, and flesh *not* blueing when handled.
7. Spores deeply colored (dark purple-brown to black).
8. Growing on or near dung.

Other Features: Stalk long in relation to width; veil thin, disappearing with age or forming a small ring that is subsequently blackened by spores; ring, if present, *not* grooved or ridged on upper surface; volva absent.

Where: Alone or in groups in dung, manure, compost, and fertilized grass; widespread and common.

Edibility: Not recommended.

Note: This is one of our most common dung mushrooms, often found with the dung bell. It resembles *Psilocybe cubensis*, the famous "magic mushroom" of the tropics, but does not stain blue when handled. There are several similar dung-lovers, including species of *Agrocybe*, which have brown spores. See MD 376 for more information.

Dung Bell (*Panaeolus campanulatus*)

Other Names: Bell-shaped Panaeolus, *Panaeolus sphinctrinus*.

Key Features:
1. Cap cone-shaped or bell-shaped, small and fragile.
2. Cap gray to brown or olive-brown.
3. Surface of cap *not* sticky or slimy and *not* netted or ridged.
4. Edge of cap adorned with small white toothlike veil remnants, at least when young.
5. Gills gray to black (*never* brown), their faces mottled with darker and lighter areas.
6. Stalk very thin ($<\frac{1}{8}$") and brittle, gray or brown (*not* white).
7. Growing on or near dung.

Other Features: Cap bald or breaking up to form a few scales; veil present, covering gills when very young but not normally forming a ring on stalk; stalk often long; volva absent; spores black.

Where: In or near dung, especially of cattle, often in small groups or colonies; widespread and very common.

Edibility: Not recommended.

Note: This dung-lover is sure to be found wherever cattle graze. If a chunk of dung laden with mushrooms is kept in a moist chamber, new ones will appear every day for several weeks — a veritable "dung garden." There are several similar *Panaeolus* species. See MD 353–360 and plate 83.

Liberty Cap (*Psilocybe semilanceata*)

Key Features:
1. Cap tiny (<1" broad), narrowly cone-shaped to bell-shaped, with a central knob or "peak."
2. Cap brownish when fresh but fading to tan or buff as it loses moisture.
3. Surface of cap sticky when moist, the skin peeling cleanly from the flesh.
4. Mature gills dark gray to purple-brown.
5. Stalk very thin (<⅛"), whitish, *without* scales.
6. Ring and volva absent (veil, if present, soon disappearing).
7. Spores deeply colored (purple-brown or darker).
8. Growing in grass.

Other Features: Cap sometimes with bluish or olive stains; sides of gills *not* prominently mottled; stalk often curved; volva absent.

Where: In tall grass, meadows, and pastures but *not* in dung or on lawns; Pacific Northwest, common.

Edibility: Hallucinogenic; it contains psilocybin and psilocin and is often used as a recreational drug.

Note: This little mushroom has several look-alikes, but the above set of fieldmarks will distinguish it. Because it likes to hide in tall sedge grass, people scouring meadows for it are forced to adopt a characteristic "psilocybe stoop" (below) that is recognizable from afar. See MD 368–371.

Potent Psilocybe
(*Psilocybe cyanescens*)

Other Names: Blueing Psilocybe, Magic Mushroom.

Key Features:
1. Cap domed (*not* sharply cone-shaped) when young, flat or often wavy at maturity.
2. Cap reddish-brown, dark brown, or caramel-brown (sometimes with bluish stains), fading to tan, dull yellow, or buff as it loses moisture.
3. Surface of cap bald, slightly sticky when moist.
4. Gills brown when young, darker in age.
5. Stalk slender, whitish when fresh but developing bluish or blue-green stains when handled or in age.
6. Veil present when young, composed of hairs or fine fibers, *not* forming an obvious ring on stalk.
7. Spores deeply colored (purple-gray to dark purple-brown), *never* dull brown, yellow-brown, or rusty-brown.

Other Features: Small to medium-small; gills typically attached to stalk; veil white, often leaving a few remnants on stalk or edge of cap; volva absent.

Where: In wood chips, landscaped areas, at the edges of woods, etc., usually in groups; common in the Pacific Northwest and northern California, and spreading.

Edibility: Strongly hallucinogenic. It contains high concentrations of psilocybin and psilocin.

Note: Like most "psilocybin mushrooms," this one stains blue or greenish-blue, but beware: not all mushrooms that stain blue contain psilocybin! A smaller species, *P. stuntzii*, is common in the Puget Sound area; it has a small ring on the stalk. See MD 368–369, 371–373, and plate 88.

Inky Cap (*Coprinus atramentarius*)

Other Names: Alcohol Inky Cap, Common Inky Cap.

Key Features:
  1. Cap 1–2″ high and broad, at first oval, then cone-shaped or bell-shaped.
  2. Cap tan to grayish-brown when young, grayer with age.
  3. Cap and gills digesting themselves from the bottom (edge of cap) upward as they mature, turning into an inky black fluid.
  4. Gills crowded together, at first white, then turning gray before becoming black and inky.
  5. Stalk white or mostly white, hollow.

Other Features: Fragile; cap bald or sometimes with a few silky white fibers or small scales, the edge striate (with fine radial lines or grooves) in age; gills free from stalk or nearly so; veil present when young, disappearing in age or forming a slight ring or small volva; spores black

Where: Grouped or clustered on ground or rotten wood in many habitats (lawns, roadsides, gardens, old fields, hardwood and mixed forests, etc.); widespread and common.

Edibility: Not recommended. If eaten with, or even several days before, alcohol, it can produce transitory but alarming symptoms: rapid heartbeat, a flushed face, metallic taste, and puffiness or numbness in the hands and feet.

Note: In urban areas this mushroom is as common as the shaggy mane, but keeps a lower profile. It is also frequent in forests with aspen, cottonwood, and birch. There are several similar, usually smaller, species of *Coprinus*. See MD 347–350 for more information.

Left: This clump of large, young "shags" is at the perfect stage for eating.
Right: Some people write with "ink" from mature shags like this one.

Marinated Shaggy Manes

Slice them lengthwise, then sprinkle with a basic vinaigrette: three to four parts extra virgin olive oil to one part vinegar. For tasting pleasure I use a simple vinegar infused with wild mushrooms that are edible but not particularly used in fine cookery. For aroma I use vinegar infused for a week with sliced, dried puffballs. If the acid is mild, say lemon, perhaps a one-to-one ratio with oil might be better. Any wonderful, complementary, minced fresh herb may be strewn over the top of the shaggy manes before serving.

—*Maya Spakowski*

Left: "Shaggy mania": the tall white caps shine brightly in headlight beams, leading to impromptu nighttime harvests such as this one.
Right: The delicate caps should be wiped clean or scraped gently with a knife, using as little water as possible.

Shaggy Mane (*Coprinus comatus*)

Other Names: Shaggie, Shag, Lawyer's Wig.

Key Features:
1. Cap tall (2–10″ high) and cylindrical when young.
2. Cap shaggy, with brown scales (at least at center) on a white background.
3. Cap and gills digesting themselves from the bottom (edge of cap) upward as they mature, turning into an inky black fluid.
4. Stalk long, white, hollow.
5. Veil present when young, forming a small, often movable ring on lower stalk.

Other Features: Medium-sized to fairly large but fragile; cap becoming grayish and broadly bell-shaped as it liquefies; gills extremely crowded (like pages in a book), white at first, black in age, free from stalk; volva absent; spores black.

Where: Usually in groups or troops along roads and paths, around parking lots and compost heaps, on lawns, and in other areas where the ground has been disturbed; widespread and common. It is partial to cool or cold weather.

Edibility: Popular, but perishable and delicate. Pick only those that have not begun to liquefy and eat them as soon as possible. Also beware of those growing along busy roads, as they may be contaminated by exhaust fumes.

Note: Shaggy manes are the soldiers among mushrooms — hundreds can be seen "marching" in columns or ranks along country roads. They will even push up through asphalt — a remarkable feat for so fragile a fungus. See photos on facing page; also, MD 345–346 and plate 85.

Sulfur Tuft *(Hypholoma fasciculare)*

Other Names: Clustered Woodlover, *Naematoloma fasciculare.*

Key Features:
1. Cap yellow, orange-yellow, or greenish-yellow.
2. Surface of cap bald, *not* sticky or slimy.
3. Gills yellow to greenish-yellow when young, becoming gray to purple-black with age.
4. Stalk slender, yellow or tawny.
5. Spores deeply colored (purple-brown to deep purple-gray).
6. Growing in tufts or clusters on rotten wood.

Other Features: Small to medium-sized; cap domed when young, becoming flatter in age; transient veil present when young, sometimes leaving remnants on edge of cap or a very slight ring on stalk; volva absent; taste usually bitter.

Where: Tufted or clustered on logs, stumps, and wood chips; widespread and common on hardwoods as well as conifers.

Edibility: Poisonous.

Note: The yellow to chartreuse color of the young gills is the most distinctive feature of this common wood-rotter. A similar species, *H. capnoides,* has grayish gills when young, a mild taste, grows only on conifers, and is edible. See MD 382-384 and plate 92.

Deadly Galerina (*Galerina autumnalis*)

Other Names: Autumn Galerina.

Key Features:
1. Cap small (<2" broad and usually ±1"), brown to tan or yellowish.
2. Surface of cap sticky when moist, bald.
3. Gills tan or brown.
4. Stalk slender (<¼"), *without* scales.
5. Veil present, at first covering the gills, soon forming a small ring on upper stalk that may disappear with age.
6. Spores brown or rusty-brown.
7. Growing on wood.

Other Features: Edge of cap striate (with fine radial lines) when moist; stalk hollow, brown in age but paler when young; ring white at first, but often brown in age; volva absent.

Where: On rotten wood or wood chips (or sometimes on buried wood), alone or in small groups or tufts but *not* in big clusters; widespread and common but easily overlooked.

Edibility: Deadly poisonous! It contains the same toxins as the deadly amanitas (death cap and destroying angel), but is seldom eaten because of its small size.

Note: This little brown mushroom is sometimes mistaken for one of the hallucinogenic psilocybes, but can be distinguished by its brown spores. There are many related species of unknown edibility — a compelling reason to avoid all little brown mushrooms unless you are absolutely certain of their identity. See MD 32–33, 401, and 892–893 for more information.

Golden Pholiota (*Pholiota aurivella*)

Other Names: Lemon-yellow Pholiota.

Key Features:
1. Cap bright yellow to golden-orange with scattered brown to reddish-brown scales or patches.
2. Surface of cap slimy or sticky when moist, shiny when dry.
3. Gills yellow to brown or rusty-brown, attached to stalk.
4. Veil present, at first covering the gills, then disappearing or leaving hairy remnants on edge of cap and sometimes forming a slight ring on stalk.
5. Stalk pale yellow to yellow, with scales below the veil.
6. Spores brown.
7. Growing on wood.

Other Features: Medium-sized; scales on cap flat to slightly raised, but not erect; stalk surface dry or slightly sticky; volva absent.

Where: On logs and stumps, or on wounds of living trees, or sometimes wood chips, usually in groups or clusters; widespread and fairly common.

Edibility: Not recommended. In addition to being slimy, it sometimes causes digestive upsets.

Note: The above fieldmarks typify a confusing group of glutinous, golden pholiotas. Some members of the group (e.g., *P. limonella*) favor hardwoods; others (e.g., *P. abietis*) grow on conifers, and it is unclear how many are distinct species. See MD 390–391 and plate 95.

Scaly Pholiota (*Pholiota squarrosa*)

Other Names: Bristly Pholiota.

Key Features:
1. Cap and stalk covered with prominent erect or curled scales.
2. Scales on cap pale tan to brown, the background usually paler.
3. Surface of cap dry, *not* sticky or slimy.
4. Stalk at least ¼″ thick.
5. Odor often garlicky or onionlike (but see note below).
6. Spores brown.
7. Growing in clusters on wood or at the bases of trees.

Other Features: Medium-sized; gills at first yellowish, tan, or slightly greenish, becoming brown in age; veil present, at first covering the gills, then sometimes forming a fragile ring on stalk; volva absent.

Where: Clustered on hardwood stumps and logs or at the bases of trees (especially aspen), sometimes also on conifers; widespread in northern latitudes and at high elevations. It is plentiful wherever aspen occurs.

Edibility: Not recommended; some are made ill by it.

Note: This attractive, brown-spored aspen-lover is easily recognized by its blatantly scaly cap and stalk. It frequently has an onionlike or garlicky odor, but an unscented form also occurs. *P. squarrosoides* (MD plate 98) of the Pacific Northwest is very similar, but its cap is sticky or slimy beneath the scales. *P. terrestris* is also similar, but grows in clusters on the ground. See MD 389–390 and plate 97.

Jumbo Gym (*Gymnopilus ventricosus*)

Other Names: Giant Gymnopilus, Big Laughing Mushroom,
Gymnopilus spectabilis, *G. junonius*.

Key Features:
1. Medium-sized to very large (mature cap 3–20" or more, stalk ½–4" thick).
2. Cap and stalk yellowish to orange, *not* sticky or slimy.
3. Gills pale yellow when young, orangish or rusty-orange in age.
4. Stalk *without* scales.
5. Veil present, at first covering the gills, then often (but not always) forming a ring on stalk.
6. Taste bitter (chew on a piece of cap, then spit it out).
7. Spores orange to rusty-orange (spore dust often visible on ring or stalk).
8. Growing on dead wood.

Other Features: Cap bald or breaking up to form small scales; flesh thick, pale yellow; gills attached to stalk but running down it very little, if at all; stalk often thickest in middle or above base; volva absent.

Where: On logs, stumps, and buried wood (especially of pine), usually in clusters; widespread and common.

Edibility: Too bitter to eat; a close relative contains psilocybin and is hallucinogenic (see below).

Note: This mushroom is very similar to the big laughing mushroom *(G. spectabilis)* of Japan and eastern North America. However, it seems to lack the inebriating qualities of the big laughing mushroom, so "jumbo gym" is a better name for it. See MD 410–411 and plate 99 (as *G. spectabilis*).

Alaskan Gold (*Phaeolepiota aurea*)

Other Names: Golden False Pholiota, *Pholiota aurea, Togaria aurea.*

Key Features:
1. Medium-sized to very large (cap 3–15" broad, stalk at least ¾" thick).
2. Cap and stalk golden-tan to golden-brown or orange-brown, covered with powder or flaky granules that are easily rubbed off (or washed off by rain).
3. Surface of cap and stalk dry, *never* sticky or slimy.
4. Gills yellowish to brown.
5. Veil present, at first funnel-shaped and covering the gills, then forming a large ring on stalk.

Other Features: Underside of ring also with granules; gills attached or free; volva absent; spores yellow-brown or ochre.

Where: In groups or large clumps on ground, usually at edges of woods or along roads and paths, especially near alder; Pacific Northwest and Rocky Mountains. Though infrequent to rare in most of its range, it is often abundant in Alaska.

Edibility: Prized by some people for its fine flavor, but poisonous to others, causing gastrointestinal distress. It should never be served to large groups of people.

Note: The large size, funnel-like veil, and granular garb make this one of the most impressive of all gilled mushrooms. Even when the granular coating is washed off the cap by rain it is usually evident on the stalk. The veil is never hairy or cobwebby as in *Cortinarius*. See MD 412–413 for more information.

Gypsy Mushroom (*Rozites caperata*)

Key Features:
1. Medium-sized (cap 2–6″ broad, stalk at least ½″ thick).
2. Cap tan to golden-brown, usually wrinkled radially, *not* sticky or slimy.
3. Surface of cap *without* scales or hairs, but often overlaid with a thin white film when young.
4. Gills pale tan to brown.
5. Veil present, *not* cobwebby or hairy, at first covering the gills, then forming a distinct ring on stalk.
6. Spores brown or rusty-brown.
7. Growing on ground in forest or tundra.

Other Features: Cap at first domed, then becoming flat; flesh white; gills usually attached to stalk; volva absent.

Where: On ground near or under conifers, in birch forests, and in tundra with dwarf birch; widespread and common in northern latitudes, southward to northern California.

Edibility: Delicious, but be sure not to confuse it with *Galerina, Agrocybe,* and *Cortinarius.* In Europe and Japan it is commonly sold in farmer's markets.

Note: The well-formed ring on the stalk, brown spores, and wrinkled cap are distinctive. So is the thin whitish film on the cap, when present. If you find a similar mushroom growing in wood chips, grass, or by a road, it is likely to be *Agrocybe.* See photo on p. 124; also see MD 412, 469–470, and plates 101–102.

Violet Cort (*Cortinarius violaceus*)

Key Features:
1. Cap and stalk deep purple.
2. Surface of cap roughened by many hairs or small scales, *not* sticky or slimy.
3. Gills deep purple when young, gradually becoming browner as they mature.
4. Stalk usually somewhat hairy or woolly, *not* sticky or slimy.
5. Veil present when young, purple, cobwebby (composed of fine fibers or hairs), soon disappearing or leaving a few hairs on stalk.
6. Spores rusty-brown.

Other Features: Medium-sized; flesh deep purple to grayish-purple; gills typically attached to stalk; stalk usually but not always thickened at base; volva absent.

Where: On ground or next to rotting logs in old coniferous forests and mixed woods; widespread but infrequent in northern latitudes, rare southward.

Edibility: Edible but hardly incredible; its principal appeal is its beauty.

Note: A memorable find! There are many other purple species of *Cortinarius* (the largest genus of mushrooms), but none as deeply colored as this one. The hairs or scales on the cap are also distinctive. See MD 417–419 and 446 for more information.

Northern Red-dye
(*Dermocybe sanguinea*)

Other Names: Blood-red Cort, *Cortinarius sanguineus*.

Key Features:
1. Entire mushroom red to dark red when fresh (but gills becoming red-brown to rusty-brown when older).
2. Surface of cap *not* sticky or slimy and *not* changing color markedly as it loses moisture.
3. Stalk *without* a bulb at base.
4. Veil present when very young, cobwebby (composed of fine fibers or hairs), soon disappearing.
5. Spores rusty-brown.

Other Features: Fairly small; cap smooth or often with small scales; gills attached to stalk; stalk usually slender, of more or less equal width throughout, *never* sticky or slimy; veil reddish; volva absent.

Where: Scattered or in groups on ground under conifers (especially spruce); fairly common from Alaska to British Columbia, south to northern California but rare.

Edibility: Not recommended.

Note: Dyers and weavers esteem this beautiful little mushroom even more than they do its yellow-stemmed cousins, *D. phoenicea* and *D. semisanguinea*. It is sometimes confused with the California red-dye, *D. californica*, a red-orange species (see photo on facing page) that gives less intense color. See MD 454–455 (as *Cortinarius sanguineus*) for more information.

The California red-dye, *Dermocybe californica* (=*Cortinarius californicus*), grows under conifers in California and the Pacific Northwest. It resembles the northern red-dye, but shows more orange in the stalk and gills. Also, its cap tends to fade as it loses moisture. The colors it gives are vivid, but not quite as intense as those of the northern red-dye.

Sweaters made from mushroom-dyed wool. The reds and oranges are derived from the northern red-dye, the blues and greens from a variety of mushrooms including the hawk wing (pp. 204–205). The sweaters were made by the woman on the right, Christina Thunholm of Sweden.

Western Red-dye
(*Dermocybe phoenicea*)

Other Names: Western Red-capped Cort, *Cortinarius phoeniceus* var. *occidentalis*.

Key Features:
1. Cap dark red to maroon or reddish-brown, *not* sticky or slimy.
2. Gills red to dark red (but becoming red-brown to rusty-brown in old age).
3. Stalk yellow or yellowish (sometimes tinged red or with reddish veil remnants, but *never* predominantly red).
4. Stalk *without* a bulb at base.
5. Veil present when very young, cobwebby (composed of fine fibers or hairs), soon disappearing.
6. Spores rusty-brown.

Other Features: Fairly small; cap *not* fading markedly as it loses moisture; gills attached to stalk; stalk usually slender, of more or less equal width throughout, *not* sticky or slimy; veil usually yellowish; volva absent.

Where: On ground in woods from the Pacific Northwest to central California. It is often abundant in pine forests with an understory of rhododendron or salal, but is difficult to see in such surroundings.

Edibility: Not recommended.

Note: This beautiful mushroom and its close relatives are prized by dyers and weavers for the beautiful shades of red, orange, rose, and purple they impart to wool. The dye pigment is chemically similar to cochineal. See MD 454 (as *Cortinarius phoeniceus*) for more information.

The eastern red-dye, *Dermocybe semisanguinea* (=*Cortinarius semisanguineus*), resembles the western red-dye but has a yellowish to yellow-brown or olive-brown cap. It is abundant in Alaska and adjacent Canada but infrequent elsewhere in western North America.

Fixin' to Dye?

If wool, why not hair? Brigid Weiler of Cortes Island, British Columbia, dyed her hair pale rose (below) by using a few dried western red-dyes. The color lasted for two months, and would undoubtedly have been more vivid had she used more (and fresher) mushrooms. The dyer's polypore, dead man's foot, jack o' lantern mushroom, and the tubes of the king bolete are also good candidates for hair-dyeing because they don't require toxic mordants to give good colors.

Poison Pie (*Hebeloma crustuliniforme*)

Key Features:
1. Cap tan to whitish, sticky or slimy when moist, bald (without fibers or hairs).
2. Gills pale tan to brown with white edges, attached to stalk and usually notched (dipping in) where they join it.
3. Stalk white, at least ¼" thick, with small white dandruffy flakes or powder at top.
4. Veil, ring, and volva absent.
5. Odor of crushed flesh radishlike.
6. Spores brown.

Other Features: Medium-sized; cap domed at first, then flat or with a broad hump at center; edge of cap rolled under at first, not normally splitting in age; stalk thickest at base or with a small bulb; flesh white.

Where: On ground in woods and at their edges, and in tree-studded lawns and cemeteries, usually in groups or troops; widespread and common. Oak and pine are favored hosts.

Edibility: Poisonous, causing moderate to severe gastrointestinal distress.

Note: This ubiquitous, attractive mushroom is, unfortunately, not good to eat. The cap color is reminiscent of a pie crust, hence its popular name. There is no veil when young as in *Cortinarius*. See MD 464–465 for more information.

Inrolled Pax (*Paxillus involutus*)

Other Names: Poison Pax.

Key Features:
1. Cap brown, the edge rolled under when young.
2. Gills running down the stalk, pale when young, becoming yellowish with age and staining brown where bruised.
3. Gills and flesh *not* exuding a juice or milk when cut.
4. Stalk whitish to yellowish but staining brown when handled.
5. Veil, ring, and volva absent.
6. Spores brown.
7. Growing on ground.

Other Features: Medium-sized; mature cap flat or with a sunken center; gills often branched near the stalk to form large pores; stalk central to slightly off-center, *not* snapping open cleanly like a piece of chalk.

Where: On ground in woods and near planted birches; widespread and common.

Edibility: Not recommended. It is poisonous raw, and people who have eaten the cooked mushroom for years can suddenly develop a serious allergy resulting in kidney failure or even death.

Note: This mushroom is reminiscent of a milk cap (*Lactarius*), but the spores are brown and no milk or juice flows when the gills are cut. The tendency to stain brown is also distinctive. See MD 477–478 for more information.

Pine Spike
(*Chroogomphus rutilus* & *C. vinicolor*)

Key Features:
1. Cap bald and slightly sticky when moist, usually shiny in dry weather.
2. Gills widely spaced and running down the stalk, orangish when young, becoming gray or black with age.
3. Flesh pale orange in cap, pale orange or yellow in stalk (but sometimes reddish throughout when old).
4. Stalk solid (*not* hollow), *without* scales.
5. Spores dark gray to black.
6. Associated with pine.

Other Features: Medium-sized; cap domed or slightly pointed when young, broader or flatter in age, color variable: grayish to orangish, red-brown, wine-red, or burgundy; transient hairy or cobwebby veil present when young but rarely forming a ring; stalk often long and curved, color variable; volva absent.

Where: On ground near or under pines, often in groups or troops with slippery jacks (*Suillus* species); widespread and common, especially in coastal California.

Edibility: Mediocre; drying improves the flavor and texture.

Note: Pine spikes are easy to recognize as a group but difficult to distinguish from each other. The two species described here differ microscopically, and others also occur. See MD 484–487 and plate 113 for more information.

Woolly Pine Spike
(*Chroogomphus tomentosus*)

Key Features:
1. Cap pale or dull orange to ochre.
2. Surface of cap dry to slightly sticky, with woolly hairs or fine scales.
3. Gills widely spaced, running down the stalk, yellow-orange to orangish when young, becoming gray to black with age.
4. Flesh pale orange in cap, pale orange or yellow in stalk.
5. Stalk solid (*not* hollow), *without* scales.
6. Spores dark gray to black.

Other Features: Medium-sized; cap at first rounded or pointed, broadening or flattening with age; transient hairy or cobwebby veil present when young, rarely forming a ring on stalk; stalk often curved, colored like cap; volva absent.

Where: On ground near or under northern conifers, often in groups or troops; very common in the Pacific Northwest, northern California, and the northern Rocky Mountains.

Edibility: Mediocre.

Note: The dry, hairy or woolly, orangish cap distinguishes this pine spike from its relatives. *C. leptocystis* is a similar species with a grayer cap. See MD 487 for more information.

Hideous Gomphidius
(*Gomphidius glutinosus*)

Other Names: Glutinous Gomphidius.

Key Features:
1. Cap dingy purple to purple-gray, reddish-brown, or sometimes brownish, often spotted or blotched with gray or black.
2. Surface of cap sticky or slimy when moist, bald.
3. Gills running down the stalk, white at first, then gray, and finally black.
4. Flesh white in cap and upper stalk, bright yellow in lower stalk.
5. Transient veil present, covering gills when young and sometimes forming an obscure ring on stalk.
6. Spores dark gray to black.

Other Features: Medium-sized; sticky skin peeling easily from cap; gills well spaced; volva absent.

Where: On ground near or under conifers (especially spruce, fir, hemlock); widespread, but most common in mountains.

Edibility: Edible; peel off the slimy skin before cooking.

Note: The slimy cap and brilliant yellow flesh in the base of the stalk are the hallmarks of this mushroom and its close relatives. *G. maculatus* is similar but lacks a veil and grows with larch. Two veiled species that favor Douglas-fir are shown on the facing page. See MD 481–483 and plate 112.

Gomphidius oregonensis differs microscopically from *G. glutinosus* and tends to have a slightly pinker or browner cap. It is common on the West Coast under Douglas fir, often in groups or clumps.

A Slick Adaptation?

The hideous gomphidius vies with the parrot mushroom, cowboy's handkerchief, and various slippery jacks for the title of "slipperiest and slimiest fungus among us." It is interesting to speculate on the function of the slime layer that coats the cap (and in some cases, the stalk) of these mushrooms, for it usually occurs on species with soft flesh. Perhaps it helps them to survive their humid environment. The slime may act like oil, repelling excess water and thereby ensuring that the crucial spore-bearing tissue (gills or sponge layer) matures before becoming waterlogged.

The rosy gomphidius (*Gomphidius subroseus*) is much more attractive than its brethren. It has a rosy cap, white to gray gills that run down the stem, a yellow stalk base, and dark gray to black spores. It is smaller than the hideous gomphidius and grows with Douglas-fir.

Left: A perfectly matched pair of orange birch boletes (p. 173). **Right**: Another matching pair: mother and daughter with *Boletus edulis*.

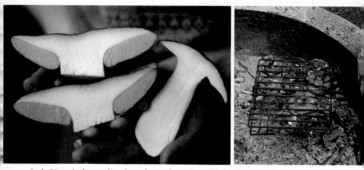

Left: King boletes sliced to show the tubes. **Right**: Roasted leccinums.

8 Boletes

Boletes have a fleshy cap and central stalk like most gilled mushrooms, but the gills are replaced by a sponge layer. The surface of the sponge is composed of hundreds of pores, but may appear smooth when very young. The pores are the mouths of closely packed spore-bearing tubes (see above photo) that constitute the sponge layer. A veil is absent in most boletes; when present it covers the sponge surface in young specimens. Most boletes grow on the ground. Polypores also have a sponge layer but are tougher and usually grow on wood (but see the blue-capped polypore, p. 190).

The king bolete and its relatives are among the most renowned of all mushrooms. A few boletes are poisonous, notably those with reddish pores. If your bolete is not among those depicted here, see MD 488–545.

Ahmed Soumille and *Boletus edulis.*

On Cooking Boletes

All boletes have a high water content which can cause them to cook up watery or slimy. Dry-sauteeing them (see p. 55) will drive off the excess moisture while concentrating their flavor. Removing the tubes (sponge layer) before cooking will also help, as they are even more watery than the cap and stalk. The tubes are full of flavor, however, and can be dried separately for use in gravies and soups (see "Essence of Edulis," p. 157). Bolete caps are also delicious basted with a little olive oil, and grilled on an open fire (see facing page).

King Bolete (*Boletus edulis*)

Other Names: Cep, Porcini, Steinpilz, etc.

Key Features:
1. Cap brown to yellow-brown, red-brown, or dark red (but often whitish when still covered by humus), bald.
2. Underside of cap with a sponge layer; pore (sponge) surface white when young, then yellowish, brown, or olive in age, *not* staining blue or brown when bruised.
3. Stalk thick (at least 1"), white to brown (never yellow).
4. Stalk surface finely netted, at least at top.
5. Flesh white, *not* staining blue or brown when cut.
6. Taste mild or nutty, *not* bitter.

Other Features: Medium-sized to large; cap sticky or dry; individual pores barely visible when white; stalk often with a bulb when young, often straighter in age; veil absent; spores dark olive-brown.

Where: On ground in woods and at their edges, often in groups; widespread and common. It favors conifers (especially pine, fir, spruce) but also occurs with oak and birch. In high mountains, the major crop is in the summer; at lower elevations, about three weeks after the first fall rains. A flush may also appear in the spring during morel season.

Edibility: One of the most sought after of all mushrooms, delectable fresh or dried.

Note: This magnificent mushroom has several forms that may be distinct species. The thick white flesh does not stain blue as in many boletes, or will at the very most blue slightly where the flesh meets the tubes. Plump as a "little pig" (*porcini*), no mushroom is more substantial or satisfying to find! See pp. 152–153 and 155–158 for more photos; also see MD 530–531 and plate 144.

This reddish-capped form of the king bolete is common in New Mexico.

Bolete Your Heart Out

It was August before I realized there was only a quart of dried boletus left on my shelf. To say panic ensued is an understatement. I was desperate. Distraught.

So I did what any desperate, distraught bolete eater would do, I went to the library and consulted the following: *One Hundred Steps to the World's Ceps; Almanac to Boletus Blooms, 1938; Hideaway Boletus Spots for the Discriminating Collector;* and *September First Is Too Late: A Guide to Boletus Bounties in the American Southwest.* Of course there were some disagreements (and more than a little vagueness) about the *exact* locations of the tastiest ceps, the most underexploited blooms, the collecting regions with the best wine and hare, but there was a general consensus that a person in my predicament should be in New Mexico in the middle of August.

A few days later I found myself in the high mountains of New Mexico, walking along the edge of a spruce forest. Aside from the fact that I was dizzy and couldn't bound around in my usual way, there was absolutely no challenge to it. My boletus stick was useless because the boletes didn't hide under the duff like they do in California. Instead they stood straight up, like statues. In an hour and a half I had filled up three containers.

My plan was to dry them all out in the morning sun. But later, alone, inching through the maggot removal and chopping-cleaning process, I confess that a new dimension of the expression "his eyes are bigger than his stomach" was revealed. As several pounds of boletes melted into oblivion I came to the realization that if I don't temper my greed, upon my death I will find myself in that portion of hell reserved for boletivores who, in punishment for their intemperance, are condemned to a limitless future cleaning an endless quantity of squirming, maggoty mushroom entrails.

Nevertheless, next August will find me in New Mexico. I'll leave the boletus stick and bring the binoculars.

—*William Rubel*

Italian-American Boletivores

Not very long ago, most of the bolete eaters in North America were immigrants from mushroom-loving countries. In California, Italian-Americans like John Feci (above right) and Enrico Re (above left) were especially prominent in the woods. Feci reminisces:

> When I was ten or twelve, I used to hunt with the older guys. They were vegetable farmers, in their forties and fifties. They couldn't speak English, but they had *class*. They weren't in a hurry. They did a lot of things by hand like make their own wine and cheese and salami and mondiola, which we'd have for lunch while they told stories of the olden days.
>
> We'd get up *early* and pile into the car like Al Capone and his gang: six guys in a Buick wearing fedoras — we didn't have the visor caps with the advertisements like they have now. The first time I went we had an automobile called a Chandler, and we couldn't make it up the hill, even in low gear. So we stopped and found a pile of boletes and coccora right there, but a lady caught us and made us give them all to her. We never found any hidden loot or bodies, but it got pretty intense. We'd see who could find the most. When we found some *cupetta* we wouldn't say "hey, come over here," like you do in a mushroom class. Instead we'd keep real quiet and try to get as many as possible before the others found out.
>
> Outsiders called us "the Italian community," but we were more like little tribes. We didn't fraternize. We'd see a lot of Italians hunting in the woods, but they didn't speak our dialect and we'd steer clear of them. The older guys all came from the same place, Campi, and they always made a big distinction between the people from Campi and the people from Pieve di Campi. Years later when I went to Italy I discovered that Campi was on a hill and Pieve di Campi was at the bottom, a couple miles away.

Enrico ("Rico") Re has hunted mushrooms for more than sixty years, since his childhood in Italy and Switzerland. Each fall he begins looking for boletes two weeks after the first inch of rain ("a half inch is no good"); he says hunting is best between the full moon and new moon. Rico likes to go "slow, slow"; he doesn't hunt to be social. He always goes alone, arriving at his favorite spot before dawn, where he waits until the sun comes up. Why does he go so early and so often? "It's what I do. When it starts raining, that's my job. When I find a big mushroom, I feel rich. I'm a king."

While more and more North Americans are discovering wild mushrooms, those who come from a long foraging tradition like John and Rico are a vanishing breed. Their children are not carrying on the tradition, perhaps because they perceive mushroom hunting as un-American. The principal measure of their success is having money, not time or knowledge. Rico's daughter won't eat the boletes he gathers, even though she knows they're delicious. "Why should I?," she asks, "when I can *buy* mushrooms at the store?"

Essence of Edulis

Sliced and dried king boletes are fantastically flavorful. Just soak them in water for about 20 minutes and then saute them for use in any dish. The tubes, however, are slimy when reconstituted, so it is better to dry them separately, soak them *overnight*, and use the liquid. Give the soaked tubes a good wringing before throwing them out (they will have lost nearly all of their flavor), then thicken the liquid with a roux (flour and butter) and season with a little salt. You now have *essence of edulis*, which you can dilute to your liking. It makes an exceptional vegetarian gravy for potatoes or rice, or can be used as a soup stock.

Slicing and drying king boletes concentrates their superb flavor.

A basket of king boletes (p. 154).

The Russian Sport of *hodit' po gribi*

[In rainy weather, the park's] shady recesses would harbor that special boletic reek which makes a Russian's nostrils dilate. . . . All alone in the drizzle, my mother, carrying a basket, would set out on a long collecting tour. Toward dinnertime, she could be seen emerging from the nebulous depths of a park alley, her small figure cloaked and hooded in greenish-brown wool, on which countless droplets of moisture made a kind of mist all around her. As she came nearer from under the dripping trees and caught sight of me, her face would show an odd, cheerless expression, which might have spelled poor luck, but which I knew was the tense, jealously contained beatitude of the successful hunter. Just before reaching me, with an abrupt, drooping movement of the arm and shoulder and a "Pouf!" of magnified exhaustion, she would let her basket sag, in order to stress its weight, its fabulous fullness . . . On a round garden table she would lay out her boletes in concentric circles to count and sort them. For a moment, before they were bundled away by a servant to a doom that did not interest her, she would stand admiring them, in a glow of quiet contentment.

—from *Speak, Memory* by Vladimir Nabokov

Queen Bolete (*Boletus aereus*)

Other Names: Moretti.

Key Features:
1. Cap dark brown when young.
2. Surface of cap with a fine whitish bloom when very young, otherwise bald.
3. Underside of cap with a sponge layer; pore (sponge) surface white when young, becoming yellowish and finally greenish as it matures, *not* staining blue or brown when bruised.
4. Stalk at least 1" thick at top, white to brown (*not* yellow).
5. Stalk surface finely netted, at least at top.
6. Flesh white, *not* staining blue or brown when cut.
7. Taste mild or nutty, *not* bitter.
8. Associated mainly with hardwoods.

Other Features: Medium-sized to large; cap often reddish-brown and/or with pale areas when mature, sticky or dry; individual pores barely visible when white; stalk often with a bulb when young, but straighter in age; veil absent; spores dark olive-brown.

Where: On ground under hardwoods (tanoak, madrone, chinquapin, manzanita, oak) or sometimes conifers. Common in California and Oregon; northern limit uncertain.

Edibility: Excellent. Some prefer it to the king bolete; others say it is bland unless dried.

Note: This species differs from the king bolete mainly in the cap color of young individuals. When older the two can be similar, although the pore surface is usually greenish in the queen bolete rather than brown as in the king. See MD 531 and plate 140.

Chuck Barrows with *Craterellus cornucopioides* (pp. 10–11).

My Most Memorable Mushroom Hunt

It was right after Paracutin erupted. There was still ash in the air, and on the blackened hillsides was a ghost forest of charred stumps completely covered with masses of sulfur shelves. It was incredibly eerie, like Halloween, as if the stumps themselves had erupted.

Another time in Mexico I found some blue-staining mushrooms in a pasture. I knew they weren't deadly poisonous, so I put them in an omelet and spent the next few hours under a blanket in the back seat of the car, laughing hysterically. My friends didn't understand. Neither did I except that I knew I was having one *helluva* time. They must have been psilocybes, but this was long before hallucinogenic mushrooms were known to science or hippies.

My most memorable recent hunt was in California. I'd wanted to taste *Craterellus* all my life, but they don't grow in New Mexico. So when I finally found some in California, at age eighty, I danced with delight. It was the last mushroom I ate for the first time, and well worth the wait — almost as good as *Boletus barrowsii*, which is the best mushroom in the world.

—*Chuck Barrows*

White King Bolete (*Boletus barrowsii*)

Key Features:
1. Cap white to pale gray or buff, bald, *not* sticky or slimy.
2. Underside of cap with a sponge layer; pore (sponge) surface white when young, then yellowish and finally olive or brownish, *not* staining blue or brown when bruised.
3. Stalk thick (at least 1″), whitish (never yellow).
4. Stalk surface finely netted, at least at top.
5. Flesh white, *not* staining blue or brown when bruised or cut.
6. Taste mild or nutty, *not* bitter.

Other Features: Medium-sized to large; individual pores barely visible when white; stalk with or without a bulb at base; veil absent; spores dark olive-brown.

Where: On ground in forests and at their edges, often in groups; restricted to drier regions (Great Basin, Southwest, California). In the Southwest it is often abundant under ponderosa pine; in coastal California it favors oak.

Edibility: Excellent — as good as the king bolete.

Note: For many years this handsome mushroom passed as a white form of the king bolete. It was finally recognized as a distinct species and named after Chuck Barrows (facing page), who pioneered the study of mushrooms in New Mexico and the Southwest. See MD 529 for more information.

This mature, handsome white king bolete weighed almost 3 lbs. It was found under an oak in coastal California.

Bitter Bolete (*Boletus rubripes*)

Other Names: Red-stemmed Bitter Bolete.

Key Features:
1. Cap tan, buff, or olive-buff, *not* sticky or slimy.
2. Underside of cap with a sponge layer; pore (sponge) surface yellow, but quickly staining blue when bruised.
3. Stalk partly or entirely red or rhubarb-colored, the upper part *not* finely netted.
4. Flesh pale yellow to whitish, staining blue when cut or bruised.
5. Taste bitter (chew on a small piece of cap, then spit it out).

Other Features: Medium-sized to fairly large; cap surface smooth or with cracks; stalk at least 1" thick, sometimes with a bulb when young, but usually straighter in age; veil absent; spores olive-brown.

Where: On ground under conifers; fairly common from the Pacific Northwest to northern California, and in the mountains to the Southwest.

Edibility: Tempting, but bitter. Cooking it doesn't help.

Note: There are several hefty, blue-staining, bitter boletes in the West, including: *B. coniferarum* of the Pacific Northwest, with a yellow to olive (*not* red) stalk that is finely netted at the top; *B. calopus*, found under mountain conifers, with at least some red or pink on its netted stalk; and *B. albidus*, with a pale cap and yellowish stalk, found under oak in California. See MD 523–525 and plate 130.

Butter Bolete (*Boletus appendiculatus*)

Key Features:
1. Cap yellow-brown, brown, or rusty-brown.
2. Surface of cap bald, *not* sticky or slimy.
3. Underside of cap with a sponge layer; pore (sponge) surface yellow, but blueing when bruised.
4. Stalk thick (>1"), yellow, at least the upper part finely netted.
5. Taste mild (*not* bitter).
6. Flesh thick and dense, pale yellow except in base of stalk, blueing erratically, if at all, when cut.
7. Flesh in base of stalk pinkish, tan, or burgundy.

Other Features: Medium-sized to large; base of stalk often (but not always) thickened or with a bulb; veil absent; spores dark olive-brown.

Where: Common on ground under hardwoods (oak, tanoak, madrone) in California; also reported from conifer forests in Oregon and Idaho.

Edibility: Edible; not nearly as sweet as the king bolete, but with wonderfully dense flesh.

Note: The large size and meaty flesh make this bolete popular with California collectors. The red-capped butter bolete, *B. regius*, is very similar but for its rose-colored cap (see MD plate 133); it occurs in California and the Pacific Northwest under both hardwoods and conifers. See MD 525–526 and plate 134.

Admirable Bolete (*Boletus mirabilis*)

Key Features:
1. Cap dark brown to maroon-brown or red-brown.
2. Surface of cap roughened by many small scales or plush-like hairs, *not* sticky or slimy.
3. Underside of cap with a sponge layer; pore (sponge) surface yellow to greenish-yellow, *not* blueing when bruised.
4. Stalk colored more or less like cap, often long (>4").
5. Base of stalk usually thicker than top.
6. Veil absent.

Other Features: Medium-sized; cap domed when young, flatter in age; flesh *not* blueing; pore surface often staining dark yellow when bruised; stalk usually netted, pitted, ridged, or streaked; spores dark olive-brown.

Where: Solitary or in small groups on or near rotting conifers, especially hemlock; common in northern California and the Pacific Northwest.

Edibility: Edible; it has a lemony flavor.

Note: This beautiful mushroom is easy to recognize. Its fondness for rotten wood is unusual for a bolete. See MD 521–522 and plates 127, 129.

> *I forgot falling off the horse*
> *With the happiness*
> *Of finding mushrooms.*
> —*Ukei*

Zeller's Bolete (*Boletus zelleri*)

Key Features:
1. Cap black to dark gray, sometimes with a reddish tinge (i.e., reddish-black).
2. Surface of cap *not* sticky or slimy, often wrinkled or uneven but *not* extensively cracked or fissured.
3. Underside of cap with a sponge layer; pore (sponge) surface yellow to olive-yellow.
4. Stalk partly or entirely red or rhubarb-colored (usually red and yellow when young and entirely red in age except for base).
5. Taste mild, *not* bitter.

Other Features: Medium-sized; flesh whitish to yellow, blueing slightly or not at all when cut and rubbed; pore surface often but not always blueing when bruised; stalk *without* a bulb, the surface *not* netted; veil absent; spores olive-brown.

Where: On ground or rotten wood in forests and at their edges; Pacific Northwest and California. It is common under both hardwoods and conifers.

Edibility: Edible.

Note: The colorful combination of dark cap, yellow pores that sometimes stain blue, and red stalk is very striking. Similar species, such as *B. chrysenteron* and *B. truncatus*, show extensive cracks or fissures on the cap at maturity, and are usually paler. See MD 518–520 and plate 128.

Slender Red-pored Bolete
(*Boletus erythropus*)

Key Features:
1. Medium-sized: cap 2–6″ broad, top of stalk ½–1½″ thick.
2. Cap brown to dark brown, sometimes shaded with red.
3. Underside of cap with a sponge layer; pore (sponge) surface red when fresh, staining blue or blue-black when bruised.
4. Stalk *without* a bulb at base.
5. Stalk yellowish, but the yellow often masked by tiny reddish dots or flakes; upper part *not* netted.
6. Flesh quickly staining blue when cut or bruised.
7. Associated with conifers.

Other Features: Cap bald or velvety, *not* sticky or slimy; pore surface sometimes orangish in old age; sponge layer yellow (except for red surface); stalk often rather long, the base sometimes with tiny velvety hairs; veil absent; spores brown to olive-brown.

Where: On ground near or under northern conifers (spruce, hemlock, fir); fairly common from Alaska to northern California, also reported from the Rocky Mountains.

Edibility: Poisonous, causing gastrointestinal distress.

Note: This beautiful mushroom is not as bulky as the other red-pored boletes, and the surface of the upper stalk is not netted. Similar species include *B. subvelutipes* of the Southwest, with a blond to yellow-brown cap and reddish to orange pore surface (see MD plates 135–136); and *B. amygdalinus*, common under oak and madrone in California, with an orange to brick-red pore surface. See MD 526–527 for more information.

Red-pored Bolete
(*Boletus pulcherrimus*)

Key Features:
1. Cap brown to reddish, often with dark fibers or small scales.
2. Underside of cap with a sponge layer; pore (sponge) surface deep red to red when fresh, staining blue or blue-black when bruised.
3. Stalk partly or completely reddish to reddish-brown.
4. Stalk surface finely netted, at least at top.
5. Stalk *without* a blatant bulb at base (but base often thicker than top).
6. Flesh quickly staining blue when bruised or cut.
7. Associated with conifers.

Other Features: Medium-sized to fairly large; cap *not* sticky or slimy; pore surface often orange or even yellowish in old age as the yellow color of the tubes (sponge layer) shows through; netting on stalk usually red or red-brown; veil absent; spores dark olive-brown.

Where: On ground near or under conifers from British Columbia to northern California, and in the Rocky Mountains and Southwest; common some years.

Edibility: Poisonous; causes severe gastrointestinal distress.

Note: The beautiful red pore surface and fine red netting on the stalk make this a memorable find. It is sometimes confused with *B. haematinus* (MD plate 138), a similar red-pored alpine species with a yellower netted stalk and brown to yellow-brown or olive-brown cap. Neither of these is as obese as satan's bolete. See MD 528 for more information.

Satan's Bolete (*Boletus satanas*)

Key Features:
1. Cap whitish, grayish, or olive-gray, sometimes with a rosy blush in old age.
2. Underside of cap with a sponge layer; pore (sponge) surface red when young and fresh, staining blue when bruised.
3. Stalk with a large bulb at base.
4. Surface of upper stalk finely netted.
5. Flesh quickly staining blue when bruised or cut.
6. Associated with oak.

Other Features: Medium-sized to large; cap bald, *not* sticky or slimy; pore surface often orange or even yellowish in old age as the yellow color of the tubes (sponge layer) shows through; bulb on stalk usually pink or rosy when young, often paler in age; netting on upper stalk usually reddish; veil absent; spores brown or olive-brown.

Where: On ground near or under oak; California and Oregon. It is sometimes abundant after the very first fall rains, a week or more before the king bolete.

Edibility: Poisonous; causes severe gastrointestinal distress.

Note: Even when the beautiful red color of the pore surface has faded, this bolete can be recognized by its briskly blueing flesh and blatantly bulbous stalk. Specimens weighing several pounds are not unusual. See MD 527–528 for more information.

Birch Bolete (*Leccinum scabrum*)

Other Names: Rough-stemmed Bolete, Brown Birch Bolete.

Key Features:
1. Cap tan, brown, or gray (in age sometimes with an olive tinge), *not* orange or reddish.
2. Edge of cap *not* rimmed with flaps of tissue.
3. Underside of cap with a sponge layer; pore (sponge) surface whitish when young, grayish to dingy brown in age, *not* blueing when bruised.
4. Stalk with numerous small projecting scales, the scales brown to black (at least at maturity).
5. Flesh white, discoloring only slightly if at all when cut and rubbed.
6. Associated with birch.

Other Features: Medium-sized; cap *not* slimy; stalk *without* a bulb, the lower part sometimes with blue or blue-green stains; veil absent; spores brown.

Where: On ground in woods or bogs near or under birch; widespread in lawns or parks where birches have been planted, and common in the wild from Washington and Montana northward.

Edibility: Edible.

Note: There are several similar edible species associated with birch. One, *L. holopus,* has a whitish cap. In another, *L. alaskanum,* the cap is mottled with dark and light areas. Still another, *L. rotundifoliae,* grows in tundra with dwarf birch. See MD 541–542 for more information.

Aspen Bolete (*Leccinum insigne*)

Other Names: Aspen Scaber Stalk.

Key Features:
1. Cap orange to orange-brown or red-brown when fresh (but often paler or duller when older).
2. Surface of cap dry or very slightly tacky to the touch, *not* sticky or slimy.
3. Underside of cap with a sponge layer; pore (sponge) surface whitish to grayish to dingy yellow-brown, *not* blueing when bruised.
4. Stalk whitish with numerous small projecting scales, the scales brown to black (at least at maturity).
5. Flesh white, but staining purple-gray to black when cut and rubbed repeatedly (especially at juncture of cap and stalk).
6. Associated with aspen.

Other Features: Medium-sized to fairly large; cap bald, the edge rimmed with thin flaps of tissue, at least when young; stalk *without* a bulb, the lower part often with blue or blue-green stains; veil absent; spores brown.

Where: On ground near aspen, often in groups or troops; abundant in northern latitudes and southward in mountains (Sierra Nevada, Rockies, etc.) wherever aspen occurs. It often fruits in June or July, before most mushrooms appear.

Edibility: Excellent; it darkens when cooked.

Note: *Leccinum* contains many edible species that closely resemble this one. In some, such as *L. aurantiacum* (found with conifers as well as aspen), the cut flesh stains reddish or burgundy before slowly darkening. See MD 540–541 and plate 147.

Panhandled Boletes

(a satyrical account)

We were driving through the Alaskan wilderness: nothing but aspen, birch, and spruce, the horizon crowned by two snowy volcanoes. There were so many mushrooms that we just *had* to stop. You should have heard us shrieking — we bolted out of the van like kids out of a classroom, not even bothering to grab our baskets and knives. I must have been seized by a *razh* — one of those mushroom passions the Russians rhapsodize about — because as I lunged from one mushroom to another, snatching them up with pagan fervor, I wasn't thinking about where I was going. *I wasn't thinking at all.*

When I finally recovered my senses, my arms were loaded with leccinums and there was an absolute quality to my surroundings that I hadn't noticed before. Absolute as in *absolute silence.* Absolute as in *absolutely no roads* for hundreds of miles except for the thin, winding gravel track we'd come in on. Absolute as in *abslutely no idea* where I was or how to find that tenuous connecting thread to civilization. Absolute as in *absolutely alone* with *absolutely nothing to eat*, not even a cheese sandwich. Wherever it was, the cramped van suddenly seemed almost comfortable, *safe,* and infinitely preferable to the prospect of hiding out from grizzlies and gnawing on raw leccinums for the rest of my life. Then my ears perked up. Was it my imagination, or were those the faint, ethereal notes of a flute drifting toward me through the aspen leaves? I couldn't believe it. I sprinted off to find the source, still tightly clutching my bolete bounty despite the thickets of leccinums looming up in my path . . .

After a frantic few minutes, I came to the edge of a glade and there, perched on a large lichened rock, was a horned, barefoot Panlike figure, nimbly playing a flute. The others stood there spellbound; like me they were laden with leccinums and had been hopelessly lost.

Then the horned figure led us, like the Pied Piper, on a merry, meandering march through that glistening, mossy, mushroom-lush wonderland. It was unbelievable — herds of hawk wings, throngs of lofty leccinums, giant fairy rings of elegant amanitas, and logs so thick with clustered caps their bark couldn't be seen. We must have looked like wood nymphs as we danced along together, completely caught up in the abundance and exuberance of the moment.

Finally he delivered us to the gravel road, where our abandoned van stood like an empty, ominous shell from some remote and irrelevant war. We offered our savior the mushrooms, which he, having a forest full of them, declined. He did allow himself to be photographed, however. Then, with an impish wink, he vanished into the forest as mysteriously as he had appeared. For a long time we stood there, straining to hear his flute, but there was nothing but the rustle of leaves in the breeze.

—*Sue Willis*

Orange Birch Bolete
(*Leccinum testaceoscabrum*)

Other Names: *Leccinum versipelle.*

Key Features:
1. Cap orange when fresh (but often fading to tan with age), *not* slimy.
2. Underside of cap with a sponge layer; pore (sponge) surface dark gray or olive-gray when young, paler in age.
3. Stalk covered with a dense layer of small projecting scales that are coal-black even when very young.
4. Associated with birch.

Other Features: Medium-sized to fairly large; edge of cap rimmed with thin flaps of tissue, at least when young; flesh darkening when cut (it may or may not stain reddish before darkening); lower stalk often with blue or blue-green stains; veil absent; spores brown.

Where: On ground near or under birch, often in large numbers; common in Alaska, south to the Rocky Mountains.

Edibility: Excellent dried or fresh, although it darkens when cooked. As with most edible mushrooms, a few people are adversely affected by it.

Note: Every summer the birch forests of Alaska and Canada swarm with these handsome boletes. In most leccinums the scabers darken gradually with age, but in this one they are black even when young, as if the stalk were dusted with coal. *L. atrostipitatum* is a very similar birch-loving bolete with a slightly duller (browner) cap. See photo at top of p. 152 and MD 541 for more information

173

Manzanita Bolete
(*Leccinum manzanitae*)

Other Names: Madrone Bolete.

Key Features:
1. Cap dark red to red-brown when fresh.
2. Surface of cap sticky or slimy when moist.
3. Underside of cap with a sponge layer; pore (sponge) surface whitish to grayish to dingy olive, *not* blueing when bruised.
4. Stalk whitish with numerous small projecting scales, the scales brown to black (at least at maturity).
5. Flesh white, darkening erratically (often slowly or only in some areas) to purple-gray or black when cut and rubbed repeatedly.
6. Associated with madrone or manzanita.

Other Features: Medium-sized to large; edge of cap rimmed with thin flaps of tissue, at least when young; stalk with or without a bulb, the lower part often with bright blue or blue-green stains; veil absent; spores brown.

Where: On ground in madrone woodlands and manzanita thickets; common from central California to southern Oregon, less frequent northward.

Edibility: Bland; not nearly as good as the aspen and birch boletes. Drying improves the flavor somewhat.

Note: In many regions where aspen and birch do not occur, this is the dominant leccinum. It is often gathered by Italian-Americans along with the king and queen boletes. Several closely related edible species also occur with madrone and manzanita, including *L. armeniacum*, which has an oranger cap. See MD 539–540 and plate 145.

Short-stemmed Slippery Jack
(*Suillus brevipes*)

Key Features:
1. Cap dark brown to reddish-brown when fresh (but sometimes fading to tan or even yellowish with age).
2. Surface of cap sticky or slimy when moist, bald, the edge *without* white veil tissue.
3. Underside of cap with a sponge layer; pore (sponge) surface white when young, pale to dull yellow in age, *not* blueing when bruised.
4. Pores *not* arranged in radiating rows.
5. Stalk usually <1" thick, of more or less equal diameter throughout, white to pale yellow.
6. Stalk surface *without* brown spots, *not* netted.
7. Veil absent in all stages.
8. Associated with pine.

Other Features: Medium-sized; flesh white to pale yellow, *not* blueing; stalk solid, often but not always short; spores olive-brown to dull cinnamon.

Where: On ground near or under pines in forests and plantations, along highways and in yards, often in masses or troops; widespread and common — perhaps the most abundant slippery jack in western North America.

Edibility: Fairly good; remove the slimy skin before cooking.

Note: The absence of a veil and lack of brown spots on the stalk distinguish this slippery jack from most of its slimy brethren. See MD 501–502 for more information

Granulated Slippery Jack
(*Suillus granulatus*)

Other Names: Dotted-stalked Slippery Jack.

Key Features:
1. Mature cap brown to cinnamon (often paler when young).
2. Surface of cap sticky or slimy when moist, bald, the edge *without* white veil tissue.
3. Underside of cap with a sponge layer; pore (sponge) surface white when young, yellowish in age, *not* blueing when bruised.
4. Pores *not* arranged in radiating rows.
5. Stalk surface with prominent resinous brown or red-brown spots (at least when mature), *not* netted.
6. Veil absent in all stages.
7. Associated with pine.

Other Features: Medium-sized; flesh white when young, pale yellow in age, *not* blueing; pore surface of young, moist specimens often beaded with milky droplets; stalk white at first, yellowish at top in age, of more or less equal width throughout; spores olive-brown to cinnamon-brown.

Where: Often abundant on ground with pines; widespread, but replaced in some regions by other slippery jacks.

Edibility: Some people are fond of it; remove the slimy skin before cooking.

Note: This widespread slippery jack has many edible look-alikes. One of these, *S. albidipes*, shows cottony white tissue on the edge of the young cap. In another, *S. punctatipes*, the pores are arranged in more or less radiating rows. Two other species are shown on the facing page. See MD 502–503 for more information.

The pungent slippery jack (*S. pungens*) of California is chameleonlike: when very young the cap is white but soon becomes grayish and then olive-gray; then as it matures it turns yellow, orange, cinnamon, or brown, or is frequently mottled with all or some of these colors. It is abundant every year under Monterey pine and other coastal pines.

Parsley, Sage, Rosemary, and Slime

Slippery jacks are so abundant that one is always looking for ways to use them. Even with their skins removed they cook up slimy. Some people get rid of the sliminess by drying and powdering them, then adding the powder to soups and gravies. Another approach is to take advantage of their inherent wetness by substituting them for escargot, as follows:

Simmer 12 small, wet slippery jack caps (preferably collected just after a rain) for 3–4 hours in a pot of water laced with finely diced carrots, shallots, salt, and a bouquet of parsley, sage, and rosemary. Take 12 *clean, empty* snail shells and stuff each one with a dollop of garlic butter and a boiled slippery jack. Place in a casserole dish with a little water and butter, and a sprinkling of fresh bread crumbs. Warm in the oven and serve.

The Kaibab slippery jack (*S. kaibabensis*) resembles *S. granulatus* but has a yellower cap; it grows in the Southwest under ponderosa pine.

Poor Man's Slippery Jack
(*Suillus tomentosus*)

Other Names: Blue-staining Slippery Jack, Woolly-capped Suillus.

Key Features:
1. Cap with gray, brown, or reddish hairs, fibers, or small scales on a yellowish to orangish background; hairs scattered to dense depending on age and weather.
2. Underside of cap with a sponge layer; pore (sponge) surface brown at first, yellowish in age (*not* white).
3. Pores *not* arranged in radiating rows.
4. Flesh and pore surface blueing when bruised or cut (sometimes slowly or only slightly).
5. Stalk yellowish to dull orange with darker spots or smears.
6. Veil absent in all stages.

Other Features: Medium-sized; cap sticky in wet weather unless covered with fibers; stalk *not* netted, the spots brown in age, resinous; spores olive-brown to dull cinnamon.

Where: On ground near or under conifers (mainly 2-needle pines); widespread and often abundant in most mountain ranges, and in coastal forests north of San Francisco.

Edibility: Insipid. In a group noted for its blandness, it ranks near the bottom. According to one collector, however, it smells and tastes like Tootsie Rolls when dried.

Note: This is our only common blue-staining slippery jack without a veil. Another "poor man's slippery jack," *S. fuscotomentosus*, has oranger pores when young and does *not* stain blue. See MD 504–505 and plates 119–120.

Slim Jack (*Suillus umbonatus*)

Other Names: Umbonate Slippery Jack.

Key Features:
1. Cap rather small (1–3″), yellowish to tan, often with an olive tinge when old.
2. Surface of cap sticky or slimy when moist, bald.
3. Underside of cap with a yellowish to olive-yellow sponge (pore) layer.
4. Individual pores large (at least 1 mm in diameter), arranged more or less in radiating rows.
5. Stalk slender (<½″ thick throughout).
6. Sticky or slimy veil present, at first covering pore surface, then usually forming a gelatinous ring on stalk.
7. Associated with pine.

Other Features: Cap often with an umbo (a low central knob); stalk solid, with obscure resinous spots that become more evident with age; no part of mushroom blueing when bruised; spores olive-brown to dull cinnamon.

Where: On ground near 2-needle pines, often abundant; southeastern Alaska to California and Colorado.

Edibility: Insipid and slimy.

Note: There are several slippery jacks (*Suillus* species) with a ring on the stalk; this one is the smallest. Others include *S. subolivaceus* (the slippery jill — see MD plate 121) and *S. sibiricus* (the Siberian slippery jack, which only sometimes has a ring — see MD plate 116), both common with 5-needle pines; and *S. luteus* (MD plate 118), usually with planted pines. See MD 498–500 for more information.

Fat Jack
(*Suillus caerulescens* & *S. ponderosus*)

Other Names: Douglas-fir Suillus.

Key Features:
1. Cap orangish to yellowish, yellow-brown, or tan (or sometimes stained dark green).
2. Surface of cap sticky when moist, bald or with a few scattered fibers.
3. Underside of cap with a sponge layer; pore (sponge) surface yellowish when fresh, *not* blueing when bruised.
4. Stalk solid within, at least the lower part blueing when cut and rubbed (sometimes slowly or weakly).
5. Stalk surface *without* resinous spots.
6. Veil present, at first covering the pore surface, then disappearing or forming a slight ring on stalk.
7. Associated with Douglas-fir.

Other Features: Medium-sized to large; pores sometimes arranged in radiating rows, staining reddish or brown where bruised, attached to stalk or running down it; veil sticky or dry, white to yellow-orange; spores brown to cinnamon.

Where: On ground in forests and at their edges, near or under Douglas-fir; often abundant on the West Coast.

Edibility: Edible.

Note: This common associate of Douglas-fir does not have the reddish-brown cap of the matte jack (*S. lakei*). The names *S. caerulescens* and *S. ponderosus* have been assigned to different extremes of what seems to be a single variable species. One striking cold-weather variant has a dark green or greenish-stained cap. See MD 496–497 and plate 117.

Tamarack Jack (*Suillus grevillei*)

Other Names: Larch Bolete, Larch Suillus, *Suillus elegans.*

Key Features:
1. Cap dark red to reddish-brown in one form, golden-yellow in another.
2. Surface of cap slimy or sticky when moist, bald.
3. Underside of cap with a sponge layer; pore (sponge) surface pale yellow to yellow or olive-yellow.
4. Stalk solid, the lower part *not* staining blue or green when cut.
5. Stalk surface *without* resinous spots.
6. Veil present, at first covering the pore surface, then usually forming a ring on stalk.
7. Associated with larch (tamarack).

Other Features: Medium-sized; underside of veil yellow and sticky or slimy before breaking; stalk pale yellow at first, reddish or reddish-brown with age; spores olive-brown to dull cinnamon-brown.

Where: On ground or moss in woods or bogs near larch, often in droves; common wherever larch occurs, from the eastern slope of the Cascades north to Alaska and east to Montana.

Edibility: Mediocre.

Note: This attractive bolete tends to have a reddish-brown cap when growing with western larch and a golden cap when growing with tamarack (eastern larch). Exceptions and intergradations occur, however. (Larch is a deciduous northern conifer whose needles grow in rosettes of 10–20.) See MD 497 for more information.

Hollow-stemmed Tamarack Jack
(*Suillus cavipes*)

Other Names: Hollow-foot, Hollow-stalked Larch Suillus, *Boletinus cavipes*.

Key Features:
1. Cap brown to reddish-brown (or occasionally yellow-brown).
2. Surface of cap covered with small fibers or hairy scales, *not* sticky or slimy.
3. Underside of cap with a sponge layer; pore (sponge) surface pale yellow, yellow, or olive-yellow.
4. Pores arranged in more or less radiating rows.
5. Lower part of stalk hollow, at least at maturity.
6. Stalk *not* blueing when cut and rubbed.
7. Veil present, at first covering the pore surface, then disappearing or forming a slight ring on stalk.
8. Associated with larch (tamarack).

Other Features: Medium-sized; veil white, often leaving remnants on edge of cap; pores usually running down stalk; stalk yellow above ring, brown below, *without* resinous spots; spores olive-brown to dull cinnamon-brown.

Where: On ground in woods and bogs near larch, usually in groups; common wherever larch occurs, from the eastern slope of the Cascades north to Alaska and east to Montana.

Edibility: Edible but hardly incredible.

Note: The cap is not slimy as in the tamarack jack, and the hollow stalk is distinctive. *Fuscoboletinus ochraceoroseus* (MD plate 123) is a similar larch-loving bolete with a pinker or redder cap and dark reddish-brown spores. See MD 494–495, 506–507, and plate 122.

Matte Jack (*Suillus lakei*)

Other Names: Lake's Bolete, Western Painted Suillus, *Boletinus lakei.*

Key Features:
1. Cap covered with reddish, reddish-brown, or brick-red fibers or small scales.
2. Underside of cap with a sponge layer; pore (sponge) surface yellow to yellow-brown or ochre when fresh.
3. Stalk solid within, staining blue or blue-green (often weakly) when cut and rubbed, especially near base.
4. Stalk surface *without* resinous spots.
5. Veil present, at first covering pore surface, then disappearing or forming a slight ring on stalk.
6. Associated with Douglas-fir.

Other Features: Medium-sized; cap dry to the touch, or sticky only beneath the hairs and scales; pores sometimes arranged in radiating rows, staining brownish or reddish where bruised; stalk yellow above ring, usually streaked with red or brown below; spores olive-brown to dull cinnamon-brown.

Where: On ground in forests and at their edges and along roads or trails, always near or under Douglas-fir; widespread and common wherever Douglas-fir occurs.

Edibility: Edible.

Note: In the Rocky Mountains and Southwest, this is the only species of *Suillus* commonly associated with Douglas-fir. In California and the Pacific Northwest it often grows with the fat jack, which has a smoother, stickier cap. See MD 495 and plate 124.

9 Polypores

These tough, shelflike or bracketlike fungi are major decayers of wood. There is a sponge layer underneath the cap as in the boletes, but in some cases the individual pores that compose it are so tiny that the surface looks smooth. Polypores can be distinguished from boletes by their shape; those that grow on the ground are tougher than boletes and usually have an off-center stalk. There is no veil.

Few polypores are tender enough to eat, but many are used medicinally (p. 193). Only a few species are depicted here. For a more detailed treatment, see MD 549–611.

Sulfur Shelf (*Laetiporus sulphureus*)

Other Names: Chicken of the Woods, *Polyporus sulphureus.*

Key Features:
1. Mushroom shelflike, growing in overlapping masses or rosettes (or sometimes singly) on logs and stumps.
2. Cap (upper surface of shelf) bright yellow-orange to orange or salmon-colored.
3. Underside of cap bright sulfur-yellow when fresh.
4. Stalk absent or nearly absent.

Other Features: Medium-sized to very large; juicy when young but tough and fibrous at maturity and fading dramatically in old age (eventually becoming brittle and whitish); pores on underside of cap often so tiny they can hardly be seen.

Where: In shelving masses or clusters or sometimes singly on logs, trunks, and stumps of both hardwoods and conifers (eucalyptus, oak, plum, fir, hemlock, spruce, etc.); widespread and common. It requires little moisture to fruit.

Edibility: Widely regarded as edible when tender, but often causing gastrointestinal distress.

Note: Also shown on the facing page, this colorful mushroom has no poisonous look-alikes but is sometimes poisonous itself. Perhaps because eucalyptus is the favored host in heavily populated central and southern California, the poisonings are often blamed on the eucalyptus. However, sulfur shelves growing on other trees have also caused digestive upsets. Conclusion: if you eat and enjoy this mushroom, *always* cook it thoroughly and do *not* serve it to lawyers, landlords, employers, policemen, pit bull owners, or others whose good will you cherish! See MD 572–573 and plates 154–155.

Beefsteak fungus sliced to show the distinctively marbled flesh.

Beefsteak Fungus Jerky

I slice them into about ⅜" strips and put them in salt water and pepper and marinate them overnight, or for 24 hours. Then I just leave them in an electric dehydrator or one of those heat things until they're real dry, like beef jerky. They taste great!

—*Chris Sterling*

Sheriff's Log

2/7 9:30 a.m. — Dorothy Willhite said one of her sons took $3,500 from her purse. The son returned the money to his mother without prompting by deputies.

2/8 11:25 a.m. — Martha Soto complained to Deputy Ingram that someone had maliciously scratched her car, causing $1,059.00 in damages.

2/8 3:35 p.m. — Bill Sanders reported a neighbor's pigs loose on his property.

2/9 8:48 p.m. — An unwanted drunk left the Boonville Lodge before a deputy could arrive to eject him.

2/12 2:17 p.m. — Bob Sanders complained to Deputy Pendergraft that a neighbor's pigs were loose on his property.

2/13 9:10 p.m. — Deputy Casella looked in vain for a domestic dispute reported to be raging on Haehl Street, Boonville.

Crime of the Week

Deputy Mason responded to reports of a woman screaming near the Branscomb Road turnoff, six miles north of Westport. The distressed woman turned out to be an exuberant mushroom hunter, screaming with delight at each new find.

—from *The Boonville Times*

Marinated raw beefsteak fungus makes a tart addition to salads.

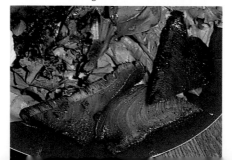

Beefsteak Fungus (*Fistulina hepatica*)

Other Names: Ox Tongue, Poor Man's Beefsteak.

Key Features:
1. Mushroom shelflike or tonguelike, exuding a dark red liquid when fresh.
2. Cap (upper surface of shelf) reddish, reddish-orange, pinkish, or liver-colored; at first velvety but becoming gelatinous in old age or wet weather.
3. Underside with a sponge layer (actually a layer of closely packed tubes or pipes).
4. Flesh streaked or marbled (best seen when sliced—see facing page).
5. Stalk absent or present only as a tough, stubby base.
6. Growing on hardwoods.

Other Features: Medium-sized to large; flesh juicy when young, tougher in age; pore (sponge) surface white to yellowish or pinkish when fresh, but aging or bruising reddish-brown.

Where: At bases of trees and on stumps, usually of chinquapin (a relative of chestnut); limited to the West Coast, most frequent in northern California.

Edibility: Edible. It has an unusual sour taste which rules out normal methods of mushroom preparation. However, it is good raw or marinated in salads and sushi, and also makes excellent jerky.

Note: This mushroom is a good source of vitamin C. When it oozes dark red juice it looks like a slab of raw meat, and when cooked it resembles liver. See MD 553–554 for more information.

Left: A mature specimen with brown cap, and a view of the small yellowish to greenish-yellow pores. **Right**: Young specimens like these are juicier and more colorful (at least at the edge) than old ones.

Dyer's Polypore
(*Phaeolus schweinitzii*)

Other Names: Red-brown Butt Rot, Schweinitz's Polypore, *Polyporus schweinitzii*.

Key Features:
1. Typically growing in rosettes with one to several caps arising from a narrowed base or short stalk.
2. Cap bright yellow-brown to rusty-brown or dark brown with an orange, yellow, or greenish-yellow edge when fresh; entirely dark brown or red-brown when old.
3. Underside of cap(s) with a shallow sponge layer; pore (sponge) surface yellow to greenish when fresh but staining brown or blackish when bruised.
4. Flesh yellowish, rusty-brown, or brown (*never* white).
5. Texture at first spongy, soon becoming tough.
6. Growing at or near the bases of conifers.

Other Features: Medium-sized to large, often engulfing needles, twigs, or other debris as it grows; very young cap woolly or hairy and yellowish to orangish; pore surface brown when old.

Where: On ground near or at the bases of conifers, or occasionally shelflike on stumps and exposed roots; widespread and very common.

Edibility: Not recommended.

Note: This mushroom attacks the roots and butts (bases) of conifers, causing them to blow over easily. It is responsible for a large amount of timber loss, yet also contributes significantly to soil fertility by breaking down the wood into usable nutrients. It is also an excellent dye mushroom (see photos). See MD 570–571 for more information.

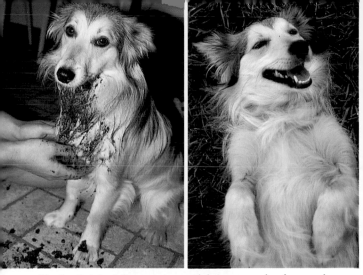

The dog that dyed: Poppy put up with being smeared and wrapped with shroom mush, then showed off her glorious new colors with pride.

My Most Memorable Mushroom Hunt

It was in the Olympic Mountains in late September. We walked up through the rainshadow and down into the rainforest. On the Quinault, the Steinpilz were looking exactly like stones on the hillside. Here would be a stone, and here a Steinpilz, just the cap. As for the Elwha, well, in those fabled days it was said the mushrooms there grew as big as rocking chairs or at least as big as golden footstools.

All day we rambled through the deep forest ways, placing mushrooms in one sack, berries and succulent salad herbs in another, with an occasional silver trout wrapped in moss and placed atop our packs, until at day's end we halted, and were amazed to see the feast spread out before us, gathered by our very own hands. The fire crackled and the stars sparkled and many smiles grew deep in our bellies and then spread to our hearts. —*David Grimes*

The dyer's polypore imparts rich yellows, oranges, golds, and browns to yarn, depending on its age and the mordant used.

Blue Knight
(Albatrellus flettii)

Other Names: Blue-capped Polypore, *Polyporus flettii.*

Key Features:
1. Cap blue, blue-green, or blue-gray when fresh.
2. Surface of cap bald, *not* sticky or slimy.
3. Underside of cap with a thin sponge layer; pore (sponge) surface white.
4. Sponge layer decurrent (running down the stalk), *not* peeling easily from the cap.
5. Stalk present, usually well developed.
6. Flesh firm and white.
7. Growing on ground.

Other Features: Medium-sized to fairly large; edge of cap often tucked under when young; cap often developing salmon or ochre stains as it ages; individual pores tiny (sometimes difficult to see); stalk central or off-center, white or tinged cap color; spores white.

Where: On ground under conifers, usually in groups or clusters; Rocky Mountains and Pacific Northwest to central California; common in some areas, otherwise infrequent.

Edibility: Edible; the firm texture requires thorough cooking.

Note: Species of *Albatrellus* are sometimes mistaken for boletes because they grow on the ground. However, they differ by their tough texture and thin, tough sponge layer which does not peel easily from the cap. There are several western species of *Albatrellus*, but only one other has a bluish cap. See MD 554-560 for more information.

Red-belted Conk (*Fomitopsis pinicola*)

Other Names: Red-belted Polypore, *Fomes pinicola.*

Key Features:
1. Mushroom shelflike or hooflike, perennial, very tough and hard.
2. Cap (upper surface of shelf) with a hard surface crust, partly or wholly reddish, cinnamon, or reddish-black (usually brown or black at base and often yellow-ochre or whitish at edge).
3. Underside white or pale yellow, *not* staining brown when scratched.
4. Stalk absent.
5. Growing on dead wood.

Other Features: Medium-sized to large; surface crust dull to slightly shiny; odor usually fragrant when fresh; pores on underside of cap often so tiny they can hardly be seen; older specimens showing stratified layers (each representing one year's growth) when chopped open.

Where: Alone or in groups on dead conifers, less frequently on living trees and rarely on hardwoods; widespread and common. It is one of the major decayers of conifers, helping to break down the wood and return nutrients to the soil.

Edibility: Much too woody to eat, but see MD 579 for a novel recipe.

Note: This common shelf fungus is sometimes mistaken for the western varnished conk, but is harder, heavier, and perennial rather than annual. See MD 578–579 for more information.

Western Varnished Conk
(*Ganoderma oregonense*)

Key Features:
1. Mushroom shelflike, with a shiny surface crust.
2. Cap (upper surface of shelf) wholly or partly reddish or mahogany (edge often yellow or ochre when young).
3. Flesh beneath the crust rather soft, punky, or corky when fresh.
4. Underside of cap white when fresh, staining brown when scratched.
5. Stalk absent, or if present then attached to side of cap.

Other Features: Medium-sized to large, annual; pores on underside tiny (sometimes not visible without hand lens), often brown in old age; stalk, when present, usually also with a shiny crust; spores brown.

Where: Occasionally found on dead or dying conifers from Alaska to northern coastal California, and in the Rocky Mountains and Sierra Nevada.

Edibility: Not edible because of its texture, but can be used medicinally as a powder, extract, or tea (see note below).

Note: The shiny surface crust, which looks varnished, is the outstanding feature of this polypore and its close relatives. A similar varnished conk, *G. lucidum* (see facing page), grows on hardwoods. The Chinese call it *ling chih* or "mushroom of immortality," because they believe it promotes longevity and good health and prevents cancer. *G. tsugae* (see photo at top of facing page) is also very similar (perhaps the same) but smaller; it grows on conifers. See MD 577–578 for more information.

Varnished conks: *Ganoderma tsugae* (left) and ling chih (*G. lucidum,* right).

Mushrooms and Medicine

I'm interested in the way cultural bias engulfs science, because scientists love to think of themselves as being free from bias. They like to think they're describing objective reality, yet they wear cultural lenses like the rest of us. In the areas of greatest emotional charge — food, sex, drugs — it's easy to see how pervasive cultural biases affect their thinking.

I like to use the example of Chinese medicinal mushrooms. In traditional Chinese medicine, drugs that have specific effects are considered the least interesting. The most highly esteemed are those with wide-ranging effects. Many mushrooms belong to this category; the Chinese believe that they strengthen the body's natural defenses and stimulate its healing mechanisms. Western medicine, on the other hand, is obsessed with finding "magic bullets," specific molecules that work on specific diseases in specific ways. If someone says a drug is good for many different conditions, Western medicine loses interest. Panaceas and tonics have the sound of snake oil.

Ginseng is a good example. For years it was something that only the "crazy" Chinese (and then hippies) used. We didn't take it seriously because its reputed wide-ranging effects didn't fit our preconceptions of medicine. When we finally got around to looking at its chemistry, we discovered that ginseng is *loaded* with interesting compounds that resemble steroid hormones and can stimulate the body's pituitary-adrenal axis (which could explain its many effects).

We know that mushrooms are full of unique, biologically active compounds like psilocybin and amanitin. Other fungi have given us some of our most powerful drugs [antibiotics], and we have one half of the world saying that mushrooms are the most desirable medicines available. Yet there is practically zero interest in them on the part of Western medicine and the pharmaceutical industry. This isn't just silly, it's completely irrational, and I believe it stems from two prejudices: one against panaceas and the other against mushrooms, a pervasive cultural belief that, beyond adding a little flavor to a dish, mushrooms are essentially worthless.

—*Andrew Weil, M.D.*

Artist's Conk
(*Ganoderma applanatum*)

Other Names: Artist's Palette, Artist's Fungus.

Key Features:
1. Mushroom shelflike, perennial, very hard.
2. Cap (upper surface of shelf) ridged and furrowed, brown to gray, *not* shiny.
3. Underside white when fresh but staining brown when scratched or bruised.
4. Stalk absent.
5. Growing on trees and stumps.

Other Features: Medium-sized to very large; pores on underside of cap so tiny they can hardly be seen; underside often brown in old age; spores brown or reddish-brown.

Where: At the bases of living trees, especially hardwoods, and on dead hardwoods or conifers; widespread and common.

Edibility: Much too woody to eat.

Note: There are many woody perennial conks, but this one is easily recognized by its white underside that stains brown immediately when scratched. As the brown staining is permanent, messages or pictures can be etched on it (see photo on p. 196), hence its popular name. It has been calculated that large specimens like the one in the above photo liberate 30 billion spores a day! Subtle air currents may lift some of the spores onto the cap, where they form a brown or red-brown powder. See MD 576–577 for more information.

The Warm Mists of October: "Of Course"

Roberto's favorite coccoli patch has been cleaned out by an honorary Italian whose ancestors wore skirts. A woodland tryst with artist Francesca is interrupted by a marauding band of ticks. Fleeing the bloodthirsty creatures, Roberto and Francesca discover the body of Erba Cahoots, socialite and naibobess. Erba's lover Lloyd suddenly appears and tries to strangle Roberto. Lloyd the campus animal control officer appears in the nick of time to save Roberto from Lloyd the mason's deadly grip. A chunk of limestone found next to the victim implicates Lloyd the mason, but Lloyd the campus animal control officer has other ideas . . .

All eyes widened in unison as Lloyd grabbed Francesca's wrist and clipped the handcuff shut. He twisted her around so that the cuffed wrist was behind her, pulled her other wrist back and, *snick*, she was in custody.

Roberto was too stunned to speak. Lloyd alertly saw his apoplexy and began his explanation.

"I'm a pro," he said. "Trained to notice small details. One clue tipped me off. But let me construct a scenario.

"Erba knows Francesca is waiting for Roberto near his favorite mushroom patch. She goes there and tells her something that is enough to make Francesca so mad she hits Erba. Maybe she doesn't mean to kill her, but that's for the judge and jury to decide.

"Then she drags the chunk of limestone over to the body to make it look like someone more athletic used it for a weapon. Someone like Erba's boyfriend Lloyd."

"You can't be serious," said Francesca. "You have no reason to believe I was near here until Roberto called me over."

Lloyd answered, "Shall I paint you a better picture? You didn't kill her with the rock because it *is* too heavy for you to swing easily. What you used was the palette you were drawing on while you waited for Roberto to show up!"

She expostulated, "But that is absurd! I *had* no palette!"

"Oh, but I believe you did!"

With that he strode to the tree and, using a handkerchief, detached the shelflike fungus that seemed to be growing there. It came away easily, having been loosely fitted back to where it had originally grown.

He turned it over, and the other two gasped to see a picture begun on the other side.

Francesca began sobbing, not very softly. "I was going to tell Roberto I was leaving him for Erba. I couldn't hide our love any longer. But she came here to say she was leaving me for Lloyd."

Two lines of tears trailed down her cheeks. "I could only have been happy with her."

The Black Maria had taken Francesca away, and the afternoon vapors began condensing into a light fog, quietly sinking to the tree tops to mantle them in a gauzy film of evanescent mist.

Roberto turned and began to leave that awful dell where his life had changed like a dark winter river carving a new channel under the pressure of raging floodwaters murky with the debris and sentiment of events back upstream.

But something bothered him. He turned back to Lloyd, still busy collecting his evidence.

"I don't believe you finished. What was the clue that tipped you off?"

Lloyd smiled. "It was a fluke, really. My train of thought was jogged when I remembered the nickname for *Ganoderma applanatum*.

"Of course! Artist's Palette."

"Actually, that thought came later. You see, another name for that hard blunt object is, of course, Artist's Conk."

The evanescent mists lingered a while around the tree tops before finally settling on the dark redwood needles, trickling together, then leaping like silver orbs of light into the void 'twixt Roberto Mortillero's shirt and neck as he pushed back to his car through the warm mists of October.

His eye was caught by the flush of motion to his right. There, darting away, was that funny little man in the skirt again! And he was acting *very* suspiciously. Against his better judgment, he turned to follow . . .   —Luen Miller

Turkey Tail (*Trametes versicolor*)

Other Names: *Coriolus versicolor, Polyporus versicolor.*

Key Features:
1. Mushroom shelflike, bracketlike, fan-shaped, or forming rosettes.
2. Texture tough: leathery and pliant when fresh, rigid when dry.
3. Cap (top of shelf or bracket) with narrow concentric zones of various colors, velvety-hairy zones usually alternating with silky-smooth ones.
4. Underside with a thin, tough sponge layer; pore (sponge) surface whitish to buff.
5. Stalk more or less absent.
6. Flesh white or pale, thin.
7. Growing on hardwoods.

Other Features: Small to medium-sized; zones on cap variable in color: usually gray, brown, and buff, but also yellow, bluish, reddish, black, white, and even green from a coating of algae; edge of cap often wavy.

Where: In groups, rows, or shelving masses on dead or occasionally living hardwoods, rarely on conifers; widespread and very common, especially in California.

Edibility: Too tough for food, but some people believe it stimulates the immune system. It can be used raw as a natural chewing gum while hiking, or taken as a tonic.

Note: This bracket fungus is a familiar feature of our oak woodlands. Several similar species occur in temperate and tropical hardwood forests; *Trichaptum abietinus*, with a hairy whitish cap and violet-tinged underside, is common on dead conifers. See MD 592–595 for more information.

Right: Leading cheers with *pom pom du blanc* (*Hericium abietis,* facing page).

Below: *Sarcodon fuscoindicum* (see MD 622) is an unpalatable deep violet tooth fungus.

10 Teeth Fungi

The teeth fungi bear their spores on downward-projecting spines or "teeth." Most teeth fungi have a cap and stalk, with the spines lining the underside of the cap. The wood-inhabiting hericiums, however, form a lovely mass of icicle-like spines without a definite cap. All teeth fungi lack a veil.

Several members of this group are delicious, and many of the tougher or more bitter ones yield beautiful dyes. Only a few species are depicted here, including one unrelated mushroom (a jelly fungus) with spines. For a more comprehensive treatment, see MD 611–630.

Bear's Head (*Hericium abietis*)

Other Names: Conifer Coral Hericium, Goat's Beard, Pom Pom du Blanc.

Key Features:
1. Mushroom branched, with many clusters of spines hanging like icicles from the branch tips.
2. Entire mushroom white, or with a yellowish or salmon tinge.
3. Growing on wood of conifers.

Other Features: Medium-sized to very large; branching compact or open; flesh white; spines mostly ¼–½" long, soft; spores white.

Where: Alone or in groups on logs and stumps of conifers; fairly common in the old conifer forests of the Pacific Northwest and northern California.

Edibility: Delicious! It can be stir-fried, marinated, or cooked like fish.

Note: This beautiful mushroom, once seen, is not easily forgotten. It grows as large as 50 pounds (but is usually less than 10) and fruits year after year from the same log. The comb hericium, *H. ramosum* (see MD plate 164), is a similar but smaller species with spines arranged in rows along the branches; it grows on dead hardwoods. See MD 614–615 for more information.

Lion's Mane (*Hericium erinaceus*)

Other Names: Old Man's Beard, Pom Pom du Blanc.

Key Features:
1. Mushroom with one large clump of spines hanging like icicles from a tough, unbranched base.
2. Spines 1–3″ long when mature, white or yellowish.
3. Growing on wood of hardwoods.

Other Features: Medium-sized to large, *without* a well-defined cap or stalk; flesh white; spores white.

Where: Growing from the wounds of living hardwoods or occasionally on dead ones; widespread. It is fairly common in California on oak, infrequent to rare elsewhere.

Edibility: Excellent, with a texture reminiscent of seafood. It is grown commercially on a small scale and served in French restaurants under the name *pom pom du blanc*.

Note: This large mushroom can appear year after year on the same tree, so when you find one, mark the spot! If you find a robust youngster, trim off a piece of it and return another day for more. See MD 615–616 for more information.

Hericiums are delicious in stir-fried dishes. The large pale chunks that look like fish are slivered lion's mane.

Toothed Jelly Fungus
(*Pseudohydnum gelatinosum*)

Other Names: False Hedgehog, White Jelly Mushroom.

Key Features:
1. Mushroom pliant and rubbery, with a cap and stalk.
2. Cap small, tongue-shaped or fan-shaped, translucent white to gray or brownish.
3. Underside of cap lined with small spines or "teeth."
4. Stalk attached to side of cap.
5. Flesh gelatinous (jellylike) or rubbery, *not* brittle.

Other Features: Stalk same color as cap or paler.

Where: On rotten wood, humus, and moss under conifers, often in groups; widespread, especially common from Alaska to northern California.

Edibility: Although bland, it can be candied or marinated or mixed with honey and cream.

Note: This pretty little mushroom behaves more like a rubber thingamajig than like a fungus: it quivers when prodded. Although it has small spines under the cap as in the teeth fungi, it is one of the jelly fungi, as evidenced by its texture. See MD 669–671 for more information.

The hedgehog mushroom ranges in color from white to pale orange, pale pinkish-tan, or orange-brown.

Hedgehog Heaven

My favorite recipe in the world is for hedgehogs. Slim Jim at the Old Milano Hotel gets 100 percent credit. He gave me a freebee one night, and *I went into mushroom ecstasy.* He prepared it with pancetta, which is an Italian bacon. It's real greasy, so he cooked it down and threw away most of the grease but kept a little to cook the hedgehogs in. Then he did a regular stir fry with the hedgehogs to get all the moisture out, the high heat business that you use for chanterelles, and he added rosemary fresh from the garden, and then pine nuts that he cooked up on the side to get them a little crispy and brownish. He put all that in with the pancetta, added some black pepper and a little bit of the grease, and then cooked it all up. Without a doubt it was the best mushroom dish I've ever had. But if you're a vegetarian or poor and pancetta costs about $7 a pound and you have to drive to Santa Rosa to get it, you can cook it without the meat. It's still delicious — you just use a little olive oil instead.

—*Chris Sterling*

The delicious bellybutton hedgehog (*Hydnum umbilicatum*), intergrades with the hedgehog but is usually smaller and slimmer, and often has a small hole or "navel" at the center of the cap.

Hedgehog Mushroom
(*Hydnum repandum*)

Other Names: Spreading Hedgehog, Sweet Tooth, *Dentinum repandum.*

Key Features:
1. Cap and stalk white to pale orange or dull orange.
2. Underside of cap with a layer of brittle spines or "teeth," white to pale orange.
3. Surface of cap bald (but may break up to form scales).
4. Flesh white and brittle, *not* woody or leathery.

Other Features: Mostly medium-sized, often with dark orange or orange-brown stains when old or bruised; cap sometimes as dark as orange-brown; spores white.

Where: On ground in forests and at their edges, often in large numbers; widespread and common. It prefers pine, spruce, and other conifers but sometimes grows with oak. It usually fruits late in the season — during the winter or spring in coastal areas with a mild climate.

Edibility: A popular edible. Many prefer it to the chanterelle, which it resembles in color and texture. When gathering it, be sure to keep the spines free of dirt.

Note: This mushroom is a blessing for the novice: it is plentiful in many regions, is usually free of maggots, and has no poisonous look-alikes. The bellybutton hedgehog (see photo on facing page) is a smaller and slimmer but equally edible species. See MD 618–619 (as *Dentinum repandum*) and plates 161–162.

A dark variety of the hawk wing (facing page) and yarn dyed with it.

Occupied by Other Tenants

Dear Mr. Arora:

I'm writing to thank you very much for your excellent mushroom book [*Mushrooms Demystified*]. Colorado, especially my side of the hill, is nowhere near as lush in mushrooms as California, but thanks to you, I have eaten well all summer and stayed out of the hospital. I guess the thing that depresses me most is that almost all of the mushrooms that I have gathered have been busy. I feel that I want to be the only one eating the thing I'm eating at the time I'm eating it. This means a great deal to me. When a mushroom is moving, and occupied by other tenants, I get very depressed. I have to throw it away. In many cases, this means letting go of the largest ones, the ones I have kicked old people and small children out of the way to obtain. You have seemed to take a rather liberal view of this, suggesting that I should just dust the maggots off. But a Colorado maggot has less to eat than a California maggot, and therefore tends to get surly when his dining pleasure is interrupted. Please advise.

—Joanne Greenberg

A wild mushroom festival is held annually in Telluride, Colorado.

Hawk Wing (*Sarcodon imbricatum*)

Other Names: Shingled Hedgehog, Scaly Hedgehog, *Hydnum imbricatum*.

Key Features:
1. Cap brown to dark brown with large brown to black scales.
2. Underside of cap with a layer of brittle spines or "teeth," brown or grayish.
3. Flesh brittle, *not* woody or leathery.
4. Tip (extreme base) of stalk *not* olive-green or black inside.

Other Features: Medium-sized to large; cap *not* lumpy or ridged, the scales often raised; taste mild or bitter; stalk white to brown; spores brown.

Where: Common on ground under spruce and other northern or mountain conifers, less common with hardwoods such as chinquapin; widespread.

Edibility: Excellent if sauteed for at least 20 minutes; otherwise it is apt to be bitter.

Note: The large, often raised scales on the cap make this mushroom look like a hawk's wing or overdone macaroon. Despite its drab color, it is an excellent dye mushroom, yielding lovely blues and greens (see facing page). Other species of *Sarcodon* also give good color, including the inedible *S. scabrosum* (MD plate 165), similar but with a black to olive-green stem base and extremely bitter taste, and several types with a smooth or cracked but not blatantly scaly cap. See MD 616–622 (as *Hydnum*) for more information.

Bleeding Hydnellum
(*Hydnellum peckii*)

Other Names: Red-juice Tooth, Strawberries and Cream, *Hydnellum diabolum*.

Key Features:
1. Beaded with red droplets when fresh and moist.
2. Cap often engulfing needles and plants as it grows (see photo at top of facing page).
3. Cap brown to blackish with a white or pinkish growing edge (but often entirely white when very young).
4. Surface of cap rough: ridged, bumpy, and/or lumpy.
5. Underside of cap with a layer of short spines or "teeth" (no sponge layer).
6. Flesh very tough and fibrous, *not* brittle.
7. Taste strongly peppery (chew on a small piece, then spit it out).
8. Stalk very tough, the flesh red-brown to dark brown when mature (*never* orange or blue).

Other Features: Medium-sized; odor often fragrant or pungent but *not* aniselike; spines pinkish to brown, the tips often paler; stalk brown; spores brown.

Where: On ground under conifers, especially pine and spruce; common from Alaska to northern California and the Rocky Mountains.

Edibility: Not edible.

Note: The red droplets on fresh, actively growing individuals are the most spectacular feature. Dyers prize this and other species of *Hydnellum* for the blues and greens they impart to yarn, especially when an alkaline dye bath is used. See MD 622–628 for more information.

Above. Mature bleeding hydnellums (without red droplets) weave a
natural tapestry. **Below**: "Old growth, new growth, woodpecker" by
Tigerlily Jones, a mushroom-dyed tapestry made from bleeding
hydnellums, hawk wings (pp. 204–205), and western red-dyes (p. 144).

Left: The purple fairy club (*Clavaria purpurea* — see MD 637) is common under northern and mountain conifers. **Above**: One of the most colorful coral fungi is the red coral mushroom, *Ramaria araiospora* (see MD 655).

Below: Pinkish or salmon-colored coral fungi are difficult to identify; this *Ramaria* dyes wool deep purple!

11 Coral and Club Fungi

These colorful fungi bear their spores on smooth to slightly wrinkled upright clubs or branches. Some are profusely branched from a fleshy base or "trunk"; others are finger-like (unbranched) but sometimes clustered. There is no veil.

Many members of this group are edible, but several are mildly poisonous and identification can be difficult. Only a few of our species are depicted here, plus two unrelated mushrooms (earth tongues) that are superficially similar. For a more comprehensive treatment, see MD 630–657.

Cauliflower Mushroom
(*Sparassis crispa*)

Other Names: *Sparassis radicata*.

Key Features:
1. Mushroom profusely branched: brainlike or resembling a sea coral or bouquet of egg noodles.
2. Entire mushroom white to slightly yellowish when fresh (but old or weathered specimens often tan).
3. Individual branches flattened (like ribbons or egg noodles), rather tough.
4. Branches arising from a tough, deeply rooting base.
5. Odor fragrant.
6. Growing at or near the bases of conifers.

Other Features: Medium-sized to very large (up to 3 ft. wide!); surfaces of branches smooth (*without* gills, pores, or spines); spores white.

Where: At or near the bases of conifers (especially pine) or sometimes on stumps; widespread and fairly common. It favors older trees or those in poor health, and will fruit year after year in the same spot.

Edibility: Prized for its fragrance and resistance to decay, but chewy and difficult to clean. Thorough cooking is necessary to make it tender; precooked, it goes well with potatoes in a casserole.

Note: This mushroom is easily recognized by its shape, color, and fragrance. It can weigh as much as 50 pounds, but 1–5 pounds is more typical. See MD 657 and plate 172.

Yellow Coral Mushroom
(*Ramaria rasilispora*)

Other Names: Northwest Spring Coral.

Key Features:
1. Mushroom profusely branched from a fleshy base or trunk.
2. Branches and branch tips bright yellow to light yellow-orange (but lower branches usually white when young).
3. Pinkish, flesh-colored, or peach tones absent.
4. Branches and trunk *not* developing wine-red stains in age or when handled.
5. Base of trunk white but *not* fuzzy.
6. Flesh inside trunk firm and white, *without* gelatinous pockets.
7. Odor and taste mild.

Other Features: Medium-sized to large; spores yellowish to ochre.

Where: On ground in woods, often half-buried in the humus; common and widespread. In the mountains it fruits during the spring as well as the summer and fall.

Edibility: Fairly good to excellent, but with a laxative effect on some people.

Note: There are many yellow coral fungi that closely resemble this one. *R. magnipes*, for instance, has a massive white base and yellow branches and is often bitter-tasting; it grows under mountain conifers in the spring and early summer. Other yellow coral mushrooms stain wine-red when bruised, or have a fuzzy white base or gelatinous flesh, or are fragrant. See MD 652–655 for more information.

Pink-tipped Coral Mushroom
(Ramaria botrytis)

Other Names: Wine-tipped Coral, Rosebud Coral.

Key Features:
1. Mushroom profusely branched from a fleshy base or trunk.
2. Base and lower branches white when young (but often tan in old age).
3. Branch tips pinkish to brick-red or wine-red when fresh.
4. Flesh inside trunk firm and white, *without* gelatinous pockets.
5. Taste of flesh mild.

Other Features: Medium-sized to large, with a large base and stubby branches when young; spores yellowish to ochre.

Where: On ground in forests under both hardwoods and conifers, often in groups or rings; common and widespread. At higher elevations it fruits in the spring as well as summer and fall.

Edibility: Edible, but with a laxative effect on some people.

Note: The above fieldmarks describe a group of handsome edible coral fungi that have collectively passed under the name *R. botrytis.* Beware of one look-alike with a very bitter or burning taste! See MD 656 for more information.

Taking hold with the hand
Of the happiness of the mountain
Mushroom gathering!
—Raisha

Flat-topped Club Coral
(*Clavariadelphus truncatus* &
C. borealis)

Other Names: Truncate Club Coral, Flat-topped Coral.

Key Features:
1. Mushroom at first clublike, but the top becoming broader and flatter with age.
2. Top of mushroom flat or shallowly concave, yellowish to orange or ochre.
3. Sides of mushroom smooth to slightly wrinkled, colored like the top or often duller (pinkish, tan, or purplish).
4. Growing under conifers.

Other Features: Medium-sized (typically 2–6" tall), unbranched or occasionally forked once, the flattened top often wrinkled; base usually whitish; taste often sweet; spores yellowish in *C. truncatus*, white in *C. borealis*.

Where: On ground near or under northern and mountain conifers, often in groups; widespread and fairly common.

Edibility: Sweet forms can be sauteed lightly and served for dessert!

Note: It remains to be determined how consistently sweet these mushrooms are, but the flattened, yellow to orange top is distinctive. The specimens in the photograph are young; their tops will broaden even more with age. Other species of *Clavariadelphus* have a rounded or pointed tip. The wrinkles on the sides of the mushroom are not as pronounced as in pig's ears (*Gomphus clavatus*). See MD 634 for more information.

Strap Coral
(*Clavariadelphus ligula* &
C. sachalinensis)

Other Names: Strap-shaped Coral.

Key Features:
1. Mushroom fingerlike or clublike, unbranched, averaging 1–4" tall and ¼–½" thick.
2. Top of mushroom rounded or bluntly pointed, *not* prominently flattened.
3. Sides of mushroom smooth to slightly wrinkled, pinkish-tan to dull yellowish or buff (*not* brightly colored).
4. Growing gregariously under conifers.

Other Features: Flesh white and stringy or pithy; taste bitter or mild; base usually whitish; spores white to pale yellow in *C. ligula*, yellowish to ochre in *C. sachalinensis*.

Where: In groups or troops on ground under conifers, especially fir, hemlock, and spruce; common from Alaska to northern California, and at higher elevations to the Southwest.

Edibility: Not worth eating.

Note: Strap corals often fruit in vast numbers, carpeting the forest floor with upright clubs, as shown in the photo. A larger species, *C. pistillaris*, is common under hardwoods in California. See MD 632–633 for more information.

Velvety Black Earth Tongue
(*Trichoglossum hirsutum*)

Key Features:
1. Mushroom small, tongue-shaped or spatula-shaped.
2. Cap and stalk black.
3. Surface of cap and stalk dry and minutely hairy or velvety, *not* sticky or slimy.
4. Texture tough but pliant.

Other Features: Stalk often twisted or curved, less than ¼" thick.

Where: In forest humus or moss, usually scattered or in groups; widespread and common.

Edibility: Too small and tough to be of value.

Note: This dainty mushroom is often overlooked because of its dark color. Several similar but less velvety black earth tongues belong to a related genus, *Geoglossum*. A few are colorful, such as the green earth tongue below. See MD 866–870 for more information.

The green earth tongue (*Microglossum viride*) is one of our few green mushrooms. It is fairly common in California's coastal redwood forests in the winter and spring.

Above: A fungo bat and puffballs. **Right**: "When it rains, it spores": mature "puff"-balls emit clouds of spore dust when pelted by raindrops, squeezed, or blown by the wind. **Below**: A woodland earthstar (*Geastrum saccatum*).

12 Puffballs and Earthstars

Puffballs bear their spores inside a round, oblong, or pear-shaped sac called the spore case. Many puffballs lack a stalk, others have a narrowed sterile base, while a few have a prominent stem. Earthstars are modified puffballs in which the spore case sits at the center of several starlike rays.

Most puffballs are edible when firm and white inside, but care must be taken to distinguish them from amanita "eggs," which show an embryo of cap, gills, and stalk when sliced open vertically (see p. 63). As puffballs mature, the interior becomes mushy or slimy and finally dry and powdery. The spore dust is disseminated by wind and rain, sometimes through a small hole that forms at the top of the puffball.

More than 100 kinds of puffballs and earthstars occur in western North America, including many prairie and desert species; only a few are depicted here. For a more comprehensive treatment, see MD 677–764.

Tumbling Puffball (*Bovista plumbea*)

Key Features:
1. Mushroom more or less round, marble-sized to golfball-sized (<2" broad).
2. At first attached to ground by a small patch of dirt-binding fibers, *without* a narrowed base or stalk.
3. Exterior *without* warts or spines, white when young but brown or lead-colored when old.
4. Interior solid and white when young, brown and powdery when old.
5. Old mushroom with a large mouth at the top.
6. Growing in grassy or open ground.

Other Features: Skin thin and papery when mature, often with a metallic luster.

Where: On ground in meadows, lawns, and other open places; widespread and common. When old it often tumbles about in the wind.

Edibility: Edible when firm and white inside, but bland.

Note: A slightly larger species, *B. pila,* is attached to the ground by a small cord rather than a patch of fibers. Many other small to medium-sized puffballs can be found in our forests, fields, and deserts. Although edible when young, they tend to be bland. Some, like *Lycoperdon* (below), have a narrowed stemlike base. See MD 690–698.

Lycoperdon perlatum is a common small to medium-sized puffball. Note the narrowed stemlike base and small hole that forms at the top in age.

Puffrut

Native Americans used puffballs as food, medicine, incense, and decoration, to staunch bleeding, to make ceremonial necklaces, and in various games. The object of puffrut (shown above and below) is to keep the puffball aloft as long as possible. It can be played with any small puffball, but for the purists only a fully mature *Bovista plumbea* (inset) will do. Although puffrut is an obvious precursor of modern games like volleyball, badminton, and hackysack, its origins are

shrouded in mystery. The derivation of the name is also unclear. "Rut" is an obvious reference to the grunting sounds of participants (reminiscent of rutting rams or seals), but does "puff" refer to the huffing and puffing of the "rutters" or the puffing of the "ball" as it is ever-so-lightly batted back and forth?

Left: For best results, keep your eye on the "ball" at all times. **Right**: Puffrutting in the early morning light.

Western Giant Puffball
(*Calvatia booniana*)

Key Features:
1. Mushroom large (at least 8" broad when fully grown).
2. Mushroom round or oblong (like a flattened sphere), *without* a prominent narrowed base or stalk.
3. Exterior with broad warts or plaques, white to pale tan.
4. Interior solid and white when young, powdery and brown or olive-brown when old.

Other Features: Skin fairly thick, eventually flaking away in plaques or disintegrating; interior with a slimy yellow intermediate phase before becoming powdery.

Where: Alone or in groups on ground in pastures and on hillsides, along roads, in sagebrush flats and other open places. Fairly common inland, but replaced on the West Coast by a slightly smaller, smoother giant puffball (see facing page).

Edibility: Edible when firm and white inside; avoid those showing traces of yellow or green, as they may cause stomach upsets. The firm, tofu-like flesh is best cubed and simmered in soups, or sliced and then breaded and fried.

Note: Giant puffballs can usually be told by their size alone. In prairie regions they can weigh as much as 50 pounds and have been mistaken by passing motorists for herds of grazing sheep. Passing mushroomists, on the other hand, are more likely to mistake herds of grazing sheep for giant puffballs! See MD 682–684 and plates 184, 186.

Above: F. Dutra and *C. booniana*. **Below**: The giant puffball (*Calvatia gigantea* or a closely related species) is common on the West Coast, especially in the spring. It is smoother and slightly smaller than the western giant puffball, and is edible when firm and white inside.

Sierran Puffball (*Calvatia sculpta*)

Other Names: Sculptured Puffball.

Key Features:
1. At least 4″ high when fully grown, broader at top than at base (but may be nearly round when young).
2. Exterior white when young, covered with large pointed warts or "peaks" resembling meringue.
3. Interior solid and white when young, brown or olive and powdery in old age (with slimy intermediate phase).
4. Growing in mountains.

Other Features: Medium-sized to fairly large; warts often pyramidal, their tips often fused to adjacent tips; skin eventually flaking away or disintegrating (small hole *not* forming at top); narrowed base sterile (*without* spore dust inside).

Where: On ground under mountain conifers and in clearings; common in the Sierra Nevada and Cascades, absent or infrequent elsewhere. It is often encountered in the spring but also fruits in the summer and fall.

Edibility: Usually good when firm and white inside, but sometimes with a slightly unpleasant, iodine-like flavor. See the western giant puffball for cooking suggestions.

Note: The spectacular warts are the hallmark of this mountain puffball. Another mountain species, *Calbovista subsculpta*, has less impressive warts but is otherwise similar. Be sure not to confuse these puffballs with warty young amanitas. The puffballs reveal solid flesh when sliced vertically, while an amanita will show an embryo of cap, gills, and stalk. See MD 684–685 for more information.

Girls Just Wanna Have Fungus

Dashing Roberto Mortillero has thrown the perfect party: Shaggy Manes, Coccoli, Blewits, Bleeding Milk Caps, Matsutake, and the Prince, each with a human escort. As often happens in California, the party ends when the host is carried out on a stretcher. The following account of the Mushroom Party is taken from the Diary of Valley Unit "Gagme" Sappa, and is reprinted with permission of the author's roommates, who are dying to see her face when she finds out it's missing.

Dear Diary: That total babe Roberto had this party tonight — I mean, if you could call it a party, like no keg! But all these babes came anyway, and I had on like this radical sequined tube top with my leopard-spotted spandex pedal pushers and high-heeled sandals, and like the babes musta thought I looked real Boss cuz they really gave me the hairy eyeball.

Like it started when I went to the hardware store cuz I read this article that all the babes go there, but like, you go up to a babe and say "Woah, that's a big pipe — I like a babe with a big pipe," and like he says he's gotta go home and paint the crib! So I see a poster for this thing called the *Fungus Fair*, and I go there, which is okay cuz there's a bunch of babes hanging out, you know like not jock types, like ee-ooo I can't stand big pecs on a guy, know what I mean? I mean like outdoorsy babes, you know, kind of like the Marlboro Man but skinnier and with dirty clothes and leaves and stuff in their hair. So I buy this mushroom book, right? It's called *Mushrooms Demystified*, like it tells how to get all the

221

"These babes, if they're munching mushrooms, they don't care what you're wearing or how much Nautilus you do or nothing . . ."

gross myst off of mushrooms, it's written by this Demystifier guy who's kinda like a Ghostbuster I guess. So anyways, I go outside and find a bunch of slimy mushrooms and go back to the Fair and ask this babe called Roberto about them, and we get talking and I say, like does he know about any good parties and he says come to his house Saturday night.

This party at Roberto's, like everyone's porking out on mushrooms, right? And *that's all*, cause these babes, if they're munching mushrooms, they don't care what you're wearing or how much Nautilus you do or nothing. And then Roberto gets like *totally sick*, and it's like gag me with a Stinkhorn, he's losing his dinner all over the place, and this babe called Norris says someone *poisoned* Roberto, and then this short fat guy in a raincoat *falls out of a tree* in Roberto's yard and lands right on top of Norris! So like the ambulance is pulling away and Gag me, there's Norris and this fat guy bleeding all over the place! Somebody says they're both dead, so it's like, Time to find another party.

So I say to this babe called Onree (I think he's like Foreign, from Oregon or someplace), if he wants to come to my place, I'll show him my Giant Puffballs. So like we step over Norris and the fat guy who fell out of the tree, and me and this babe Onree go to my place and wouldn't you know it, those sub-normal *cows* I live with, they won't leave us alone for a second, and we end up *talking*! Like Gag me, like I wanna spend Saturday night *talking*! . . .

—edited by Mike O'Ryzal, with Monty Ray Pyne

Earthstar (*Astraeus hygrometricus*)

Other Names: Hygroscopic Earthstar, Barometer Earthstar.

Key Features:
1. Mushroom up to 3″ broad when expanded, consisting of a round to somewhat flattened spore case seated at the center of 6–16 starlike rays.
2. Rays hygroscopic (folding over the spore case and completely hiding it in dry weather, then unfolding to expose the spore case when moistened).
3. Open rays *not* standing on their tips.
4. Spore case *not* mounted on a short stalk.
5. Surface of spore case roughened by fine particles.

Other Features: Rays very tough, their surfaces often cracked; spore case eventually rupturing through a hole or slit at the top and then disintegrating.

Where: On ground in deserts, roadside ditches, pastures, under shrubs or trees, etc.; widespread and common.

Edibility: Not edible when mature. In Thailand, the underground, truffle-like "eggs" are considered a delicacy.

Note: Many earthstars cannot close once open. *A. pteridis* is a large western version of *A. hygrometricus*. See MD 669-706 for more information.

A constellation of earthstars. These species lack the "earthstartling" ability to open and close at will, but are notable for standing on the tips of their rays: *Geastrum fornicatum* (three on left), *G. pectinatum* (the "beaked" one on the right), and *G. quadrifidum* (small, foreground).

Dead Man's Foot
(*Pisolithus tinctorius*)

Other Names: Dyemaker's False Puffball, Stone Puffball, Bohemian Truffle, *Pisolithus arrhizus, Pisolithus arenarius.*

Key Features:
1. Mature mushroom brown, brittle, and dusty, protruding from the ground like a dusty stump, half-rotted root, or ball of dried-up dung.
2. Young mushroom roundish to pear- or club-shaped.
3. Interior of young mushroom containing many small seedlike capsules imbedded in a dark sticky substance (slice it vertically as shown at top of facing page).
4. Seedlike capsules disintegrating into brown powder with age, the disintegration process beginning at the top of the mushroom and progressing downward.

Other Features: Medium-sized to large; exterior yellowish, brownish, purplish, or blackish; base of mushroom often narrowed like a stalk; volva absent.

Where: In disturbed or exposed soil near trees and shrubs, along roads and trails, in old pastures, etc.; widespread, but especially common in Oregon and California. It often bursts through pavement.

Edibility: Not edible.

Note: This bizarre mushroom forms a symbiotic relationship with the rootlets of various trees and shrubs, enabling the trees to survive in poor soil by helping them absorb nutrients. It is also an excellent dye mushroom, yielding gold, brown, and black. See MD 712–713 for more information.

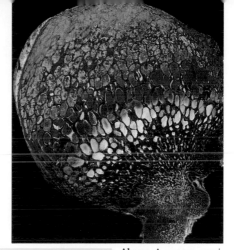

Pick-A-Lithus

Every September a pesky *Pisolithus* pokes up in the patio shown below, displacing a different brick each time (right) as its mycelium moves around in search of nutrients. Inspired by the surfeit of government-sponsored lotteries, the local mushroom club decided to hold a "fung raiser" called "Pick-A-Lithus" (*lithus*, stone) in which people bought the brick or bricks that they thought would be hoisted by the *Pisolithus* the following year. All the bricks were sold, and the winner received a "cord" of dried morels.

Above: A young specimen sliced open to show the spore-bearing capsules inside.

An expanded specimen is at center; the young one at lower left has been sliced open (right) to show the thick, hard skin and dark interior.

Dead Man's Hand
(*Scleroderma geaster*)

Other Names: Earthstar Scleroderma, *Scleroderma polyrhizon.*

Key Features:
1. Young mushroom resembling a hard puffball (i.e., more or less round) but with a very thick (⅛–⅜") skin that splits into several rays or "fingers" when mature.
2. Mushroom more than 4" broad once the rays spread out.
3. Interior hard and white when very young but becoming gray to purple-black while still firm, finally brown and powdery after the skin splits open.
4. Growing in disturbed or exposed soil, sand, or asphalt.

Other Features: Exterior smooth or slightly scaly but without warts, whitish to yellowish, usually dirty or sandy; inner surface (upper surface of rays) brown to blackish.

Where: Usually half-buried in sand, asphalt, or poor soil along roads or in other disturbed areas; widespread, especially common in California.

Edibility: Poisonous, causing gastrointestinal distress.

Note: After splitting into rays, this mushroom might remind someone with a morbid imagination of the outstretched palm of a buried corpse, especially when found near a dead man's foot, as is often the case. *S. texense* is a similar but scalier species. Other sclerodermas are smaller or do not split open so dramatically. They are easily mistaken for edible puffballs, but have a thick, hard skin and firm, purplish to black interior. See MD 707–711 and plate 190.

Desert Shaggy Mane
(*Podaxis pistillaris*)

Key Features:
1. Mushroom with a cap and stalk.
2. Cap oval or tall and narrow, white to tan.
3. Stalk tough, extending all the way through the cap to the top.
4. Gills absent; interior of cap with irregular plates when young, eventually disintegrating into reddish-brown to blackish-brown spore powder.
5. Volva absent (i.e., no sack or cup at base of stalk).
6. Growing in deserts and other dry places.

Other Features: Medium-sized; cap scaly or smooth, eventually tearing or splitting lengthwise; interior at first white, soon yellowish, then reddish, and finally darker; stalk smooth or scaly, white to tan, usually long.

Where: On ground in deserts, dry mountains, sandy waste places, etc., throughout the arid regions of the world. It appears soon after it rains but persists for weeks without decaying.

Edibility: Edible when young; in India it is highly prized.

Note: This is the most common of several "stalked puffballs" that occur in our deserts. From a distance it resembles the shaggy mane, but is much tougher, does not have gills, and does not digest itself at maturity. For more information on this and other desert species, see MD 713–723, 725–726, 727–732, and plate 187.

Morels are easily recognized by their hollow, honeycombed cap that is intergrown with the stalk along its full length.

13 Morels and False Morels

In these mushrooms a cap and stalk are present, there is no veil or volva, and the spores are borne on the exposed surfaces of the cap. The prized morels have a hollow, honeycombed cap that is completely intergrown with the stalk. In the false morels and elfin saddles (pp. 234–238), the cap is lobed, brainlike, or saddle-shaped but not honeycombed.

To mushroom hunters who live in regions with cold winters, morels are synonymous with the spring. They develop slowly, taking several weeks to attain full size. As a result, the morel season can span two months or more, moving up in elevation as the days get warmer.

Morels occur in many habitats but are often difficult to see. Their mycelia tend to be short-lived, so new "patches" must be found every year. These factors combine to make morel hunting especially challenging and competitive, hence this book's noticeable lack of specific information on their whereabouts. Those "in the know" will obviously appreciate this, but so should beginners, who surely don't want to be deprived of the "demorelizing" experience of tramping through the forest for hours, weeks, months, or even years without finding a single one. Only by being skunked repeatedly can you savor the sweetness of ultimate success!

Some of the false morels are dangerously poisonous, especially if eaten raw. Those that are edible, as well as the true morels, should always be cooked before being eaten. If your "morel" has a sack at the base of the stalk, you have a stinkhorn (see p. 241). See MD 783–816 for more information on morels and false morels.

Believing is Seeing

Mushroom hunting can teach us a lot about the larger world. A common experience of mushroom hunters is not being able to see a particular mushroom when they first try to collect it. It's not a question of visual acuity, but of pattern recognition. One woman wanted to find morels. She'd been told they grew in her area, but nobody would show her exactly *where*, and she had never seen one in the flesh. So finally she went out by herself to the woods and spent an entire morning looking, without finding a single morel. In frustration she got down on her hands and knees and began sifting through last year's leaves. Just as she was about to give up, she saw one morel a few inches away, and picked it. Clutching it triumphantly, she looked up and saw *hundreds* of them scattered through the woods in all directions.

A useful lesson can be drawn from this: that our brain acts as a filter, screening out what it doesn't consider significant. A certain "key" has to be in place before our brain can say "Aha!" and recognize something. And of course, what we recognize has real consequences. In this case, the person who can see the morels gets to put them in the basket and take them home to eat. The larger principle is that what we experience is determined by what we are able to perceive. It leads me to believe that we should be willing to accept other people's experiences —for instance, of telepathy or precognition — or at least *consider* that they have validity, even though we do not share them. Otherwise we could live in a forest full of morels and never see them.

—*Andrew Weil*

Morels can be extremely difficult to see. You should be able to spot at least six black morels (p. 230) in this photograph.

Black Morel
(*Morchella elata* & *M. angusticeps*)

Other Names: *Morchella conica.*

Key Features:
1. Cap round to cone-shaped or like a Christmas tree, honeycombed with pits and ridges.
2. Pits fairly dark (olive-brown, dark gray, black, or reddish-brown), at least when mature.
3. Ridges colored like the pits or darker (sometimes black).
4. Cap completely intergrown with the stalk (joined to it along its full length).
5. Entire mushroom hollow.
6. Stalk *without* a sack or cup at base.

Other Features: Medium-sized to large; stalk white or tinged gray or, in one variety, pinkish or reddish, often with small warts; odor *not* obnoxious.

Where: On ground in many habitats, but especially under mountain conifers in recently logged or burned areas; widespread and sometimes abundant, especially in the spring.

Edibility: Among the most highly prized of all mushrooms, delicious fresh or dried. It should always be cooked.

Note: Black morels blend so well with their surroundings that, without an accurate "search image," seekers quickly become "demorelized" (see p. 229). There is a great deal of variation in size, shape, pattern, and color, as the photos show. We don't know whether these varieties are distinct species, but we do know that they're all edible! See MD 790–791 and plates 199, 202.

A small, cone-shaped, dark-ridged variety of the black morel.

My Most Memorable Morel Hunt

We'd been scouring the rugged, rocky slopes of a burn for several hours. Some of us had struck it rich, others had just struck out. It was hard work, but there was this *tribal* feeling, so when we got back to the cars we didn't do the usual thing and each drive away with what we'd found. Instead, we spread all the morels out on towels. Then we made a tight circle around them and went round and round, taking turns choosing them one by one until all 217 had been divied up. It was memorable because we got to really examine, admire, and *covet* all those morels, and even more memorable because *everyone* scored. It was a hunt with a happy ending and a "morel": the acquisition of a bounty through *sharing.*   —*Grainger Hunt*

Bobbie Stanfield of Wenatchee, Washington with overgrown (but still edible) black morels.

Morel (*Morchella esculenta*)

Other Names: Yellow Morel, Merkle, Sponge Mushroom.

Key Features:
1. Cap round to oval or cone-shaped and honeycombed with pits and ridges.
2. Pits and ridges buff, tan, or yellow-brown at maturity.
3. Cap completely intergrown with the stalk (joined to it along its full length).
4. Entire mushroom hollow.
5. Stalk *without* a sack or cup at base.

Other Features: Mostly medium-sized; pits sometimes darker when young; stalk usually whitish; odor *not* obnoxious.

Where: On ground in many habitats: woods, fruit orchards, gardens, sandy soil, wood chips, landscaped areas, etc.; widespread. It appears slightly earlier than the black morel and is not as prevalent at high elevations. A good place to check is under cottonwood and alder along streams.

Edibility: Among the most highly prized of all mushrooms, delicious fresh or dried. It should always be cooked.

Note: This is the familiar "sponge mushroom" of the Midwest. It is fairly common in the river valleys of the West but, like the black morel, is extremely difficult to pick out against a backdrop of dead leaves. A giant version, *M. crassipes*, also occurs, especially in Oregon, and the white morel (facing page) is common in coastal California. See MD 787–789.

The white morel (*Morchella deliciosa*) is rather small with white ridges and dark, elongated pits but becomes tan overall (as shown here) with age.

Stuffed Morels

Soak one dozen dried morels in ¾ cup cream and 1 T. dry sherry. Saute a large minced shallot, a large clove of garlic, and ½ cup finely chopped pecans in 3 T. butter. After they're cooked, turn off heat and mix in 4 oz. goat cheese and a little fresh grated nutmeg and white pepper, and stuff the soaked morels. Bake 20 minutes at 325° and serve over fresh pasta.

—*David Canright*

False Morel (*Gyromitra esculenta*)

Other Names: Brain Mushroom, Lorchel, Beefsteak Morel.

Key Features:

1. Cap lobed and brainlike, the surface strongly wrinkled at maturity but *not* honeycombed with pits.
2. Cap red-brown, brown, or yellow-brown, usually attached to stalk at several points but *not* completely intergrown with it.
3. Stalk smooth or slightly grooved, *without* prominent ridges.
4. Stalk interior *not* intricately folded: cross-section of upper stalk showing only one or at most two hollows.
5. Growing in the spring.

Other Features: Medium-sized; cap surface often smooth when young; stalk not nearly as wide as cap; stalk white, reddish, or tinged cap color.

Where: On ground or rotten wood in forests, clearings, and along roads; widespread and common in northern latitudes, southward in mountains. It fruits in the spring and early summer, sometimes in the company of true morels.

Edibility: Not recommended. Some people eat it dried or parboiled, but it is deadly poisonous raw (see facing page) and may contain carcinogens even after being cooked.

Note: The brainlike brown cap is the outstanding feature of this mushroom. Another poisonous species, *G. infula* (see photos at bottom of facing page) has a smooth, not wrinkled, cap that is often saddle-shaped. See MD 801–803 and plate 206.

234

Gyromitra esculenta can be yellow-brown, brown, or liver-colored.

A Volatile Subject

Many fungophobes (including some mushroom "experts") delight in issuing dire warnings about mushrooms that are edible under some circumstances and deadly poisonous under others. These warnings, however, apply only to the false morels, a group that is easily identified.

Several false morels (e.g., *G. esculenta*) are highly toxic raw. The toxin (gyromitrin or its byproduct MMH, a rocket fuel) is extremely volatile and can be eliminated by drying the mushrooms or cooking them in an open pan (but don't inhale the vapors!). Although some people avidly eat false morels, the difference between a "safe" dose and a lethal one is very small, leading Captain Charles McIlvaine, a "toadstool-tester" renowned for his fearlessness, to say: "It is not probable that in our great food-giving country anyone will be narrowed to *G. esculenta* for a meal. Until such emergency arrives, the species would be better left alone." Certain related mushrooms such as the early morels (*Verpa*) may contain traces of MMH, so it is a good policy to cook *all* edible morels, false morels, elfin saddles, and cup fungi before eating them.

The hooded false morel (*Gyromitra infula*) has a smooth, sometimes saddle-shaped cap and grows in the summer and fall (or winter on the coast) rather than during morel season; it is poisonous, at least raw.

Snowbank False Morel
(*Gyromitra gigas*)

Other Names: Snow Mushroom, Walnut, Bull's Nose, *Neogyromitra gigas*.

Key Features:
1. Stature chunky, the stalk as thick or nearly as thick as the cap.
2. Cap brainlike or wrinkled but *not* honeycombed with pits, the lobes *not* projecting outward much.
3. Cap yellow-brown to tan when fresh (but often red-brown or darker brown in age).
4. Stalk ribbed and folded; cross-section showing complex internal folds and chambers.
5. Growing in spring or early summer.

Other Features: Medium-sized to large; edge of cap attached to stalk; stalk whitish.

Where: On ground or rotten wood, usually under conifers near melting snow or soon after snow melts; common during spring and early summer in the Pacific Northwest, and southward in mountains (Rockies, Sierra Nevada, etc.).

Edibility: Edible and popular. Although it does not have the toxic reputation of other false morels, it should always be cooked as a precautionary measure. Some people prefer it to the true morels.

Note: The thick, folded stalk and chunky stature of this mushroom distinguish it from other false morels. See MD 800–801 for more information.

Umbrella False Morel
(*Gyromitra californica*)

Other Names: California Elfin Saddle, *Helvella californica.*

Key Features:
1. Cap domed, umbrella-like, or wavy and spreading, the edge *not* attached to stalk.
2. Cap yellow-brown to dark brown, smooth to somewhat wrinkled but *not* honeycombed with pits.
3. Stalk with prominent ribs that extend up underside of cap nearly to edge.
4. Stalk stained reddish, pinkish, or burgundy, at least at base.

Other Features: Medium-sized to large; cap thin and fragile, much wider than stalk once it has unfurled.

Where: On ground in woods (especially under conifers) and at their edges, along logging roads, or sometimes on rotten wood; Pacific Northwest and Rocky Mountains, southward in mountains. Found in spring, summer, and fall, but infrequent.

Edibility: Not recommended. It is poisonous, at least raw.

Note: The thin, ribbed cap of this false morel often appears puffed up or umbrella-like. Occasionally the stalk lacks the distinctive rosy stains. Gigantic specimens are sometimes encountered. See MD 804 for more information.

Fluted Black Elfin Saddle
(*Helvella lacunosa*)

Other Names: Black Elfin Saddle.

Key Features:
1. Cap gray to black.
2. Cap lobed, the surface smooth or wrinkled but *not* honeycombed with pits.
3. Stalk white to dark gray or sometimes black.
4. Stalk with prominent grooves and ridges that branch to form elongated holes or pockets.
5. Cross-section of stalk showing complex internal folds and chambers.
6. Stalk white to dark gray or sometimes black.

Other Features: Medium-sized; edge of cap usually attached to stalk in several places.

Where: On ground in woods and under trees; widespread and common. It is especially abundant in coastal California pine forests.

Edibility: Edible when cooked. Some people fry it like a morel; others dry and powder it for use as a seasoning.

Note: This is the most common of several elfin saddles; others are differently colored or lack the fluted and pitted stalk. Like many of its relatives, this species will often shoot off a cloud of spore "smoke" when handled or otherwise disturbed (see photo of crown cup on p. 240). See MD 815–816 for more information.

Left: Early morel (*Verpa bohemica*). Right: Thimble morel (*V. conica*).

Early Morel (*Verpa bohemica*)

Other Names: Wrinkled Thimble Morel, *Ptychoverpa bohemica*.

Key Features:
1. Cap sitting on the stalk like a thimble, i.e., attached to stalk only at very top with the sides hanging free like a skirt.
2. Cap *without* strongly projecting lobes, *not* saddle-shaped.
3. Cap brown to yellow-brown and strongly wrinkled vertically, the wrinkles sometimes branched to form pits.
4. Stalk fragile and hollow, the hollow usually stuffed with cottony material.
5. Stalk *without* a sack or cup at base.
6. Growing in the spring.

Other Features: Medium-sized; edge of cap often slightly upturned in old age; stalk relatively long when mature.

Where: On ground in woods and thickets, especially in sandy soil along streams; widespread, but most common in the Pacific Northwest. The main crop is typically one to three weeks before the true morels (*Morchella*).

Edibility: Not recommended. The flavor is not as good as that of the true morels, and it can be poisonous if eaten often or in large amounts (see p. 235).

Note: This mushroom is intermediate in appearance between the true morels (*Morchella*) and the false morels (*Gyromitra*). The thimble morel (*Verpa conica*—see top right photo) is similar but has a smooth to slightly wrinkled cap. See MD 793–795 for more information.

239

Left: Cup fungi such as this crown cup (p. 246) often "smoke" (discharge spores) when handled. **Right**: The apricot jelly mushroom (*Phlogiotis helvelloides*) is pink to red-orange and funnel-shaped to spatula-shaped, with a rubbery or gelatinous texture. It can be eaten raw in salads or pickled and candied (see MD 672).

The scarlet cup (*Sarcoscypha coccinea*) is one of our most colorful cup fungi; it grows on dead hardwood sticks and logs (see MD 836).

14 Other Mushrooms

This motley assortment of mushrooms includes representatives of the cup fungi, jelly fungi, stinkhorns, truffles, flask fungi, and smuts.

The cup fungi are cuplike or disclike with spores produced on their upper or inner surface. Jelly fungi come in various shapes but are always gelatinous (jellylike) or rubbery. Stinkhorns are variously and sometimes spectacularly constructed; all have an obnoxious odor and a sack or volva at the base. Truffles produce their spores internally; they grow underground and look like small potatoes. Flask fungi and smuts are often parasitic; those depicted here grow on mushrooms and corn, respectively. For more details on these fungi, see MD 669–675, 739–781, and 817–887.

Left: A stinkhorn "egg" sliced in half. Right: A fully expanded individual.

Stinkhorn
(*Phallus impudicus* & *P. hadriani*)

Key Features:

1. Mature mushroom phallic, with a stalk and "head."
2. "Head" covered with olive to brownish-black spore slime; surface white and pitted, ridged, or netted beneath the slime (or when slime is gone).
3. Stalk white, fragile, hollow.
4. Base of stalk sheathed by a loose sack (the remains of the "egg" that initially surrounds the mushroom).
5. Odor obnoxious.

Other Features: Medium-sized; beginning as an underground "egg" with cord(s) attached; exterior of "egg" white in *P. impudicus*, pink to purple in *P. hadriani*; interior of "egg" with olive spore mass surrounded by gel (see photo).

Where: On ground or rotten wood in lawns, gardens, sandy soil, roadsides, in front of cathedrals, etc.; common in California and the Southwest, infrequent northward.

Edibility: Edible in the egg stage, before the odor develops. In central Europe and China the "eggs" are a delicacy.

Note: Stinkhorns are distantly related to puffballs. After "hatching," they elongate rapidly, thrusting the foul-smelling spore slime into a position where flies can find it. The flies feast on the slime, then carry the spores to new localities. The phallic appearance of some stinkhorns has given rise to numerous folk beliefs, and was the cause of considerable consternation in Victorian England. Some stinkhorns, however, are branched, latticed, or starfishlike; all begin as "eggs." See MD 764–778 and plates 191–198.

Witch's Butter (*Tremella mesenterica*)

Other Names: *Tremella lutescens.*

Key Features:
1. Mushroom gelatinous (jellylike), yellow to orange.
2. Shape variable: usually lobed and wrinkled or brainlike, but bloblike if swollen with water.
3. Stalk absent.
4. Growing on dead hardwoods.

Other Features: Small to medium-sized; hard when dried out.

Where: On logs, stumps, fallen branches, and sticks of oak and other hardwoods; widespread and common, especially in California.

Edibility: Harmless but flavorless.

Note: This mushroom shrivels up in dry weather, then swells with water as soon as it rains. It often shares logs with a small bracket fungus (*Stereum*). *Dacrymyces palmatus* is a similar orange jelly fungus that favors dead conifers and differs microscopically. Brown and black species also occur. See MD 672–674 and plate 170.

> *Mushroom hunting;*
> *Tall people*
> *Are no good at it.*
> *—Yayu*

Wood Ear (*Auricularia auricula*)

Other Names: Tree Ear, Judas' Ear, *Hirneola auricula-judae.*

Key Features:
1. Mushroom pliant and rubbery, shaped more or less like an ear or a shallow inverted cup.
2. Interior surface tan, brown, red-brown, or purple-brown, bald.
3. Exterior usually veined or ribbed, finely hairy, some shade of brown.
4. Flesh rubbery or gelatinous (jellylike), *not* brittle.
5. Stalk absent.
6. Growing on wood.

Other Features: Small to medium-sized; hard when dried out.

Where: On rotting wood, usually in groups; widespread, especially common in mountains.

Edibility: Edible. A similar species, *A. polytricha* (the cloud ear or tree ear), is cultivated in Asia for use in soups and can be purchased dried in Asian food stores. Both species are used medicinally in China and have been shown to reduce cholesterol buildup in the blood.

Note: Because of its shape, the wood ear can be mistaken for a cup fungus. However, its texture is rubbery rather than brittle, and there are also microscopic differences. See MD 675 for more information.

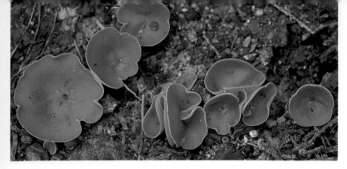

Orange Peel Fungus (*Aleuria aurantia*)

Other Names: Golden Fairy Cup.

Key Features:
1. Mushroom cup-shaped or wavy to nearly flat, usually ½–4" broad.
2. Upper (inner) surface bright orange to yellow-orange.
3. Underside paler or whitish, *without* dark hairs.
4. Mushroom *without* blue or greenish stains.
5. Stalk absent or practically absent.
6. Growing on ground, usually lying flat (*not* on end).

Other Features: Upper surface bald; flesh fragile.

Where: Scattered or clustered on soil, often along roads or paths; also in sand, moss, or grass but *not* on dung or in ashes; widespread and common.

Edibility: Edible.

Note: This little cup fungus looks like a piece of discarded orange peel but is more fragile and not as common. When tightly clustered the little cups are often distorted by mutual pressure, so that they look like the petals of a flower. A similar stalked species, *A. rhenana*, is less common. A bluish-staining species, *Caloscypha fulgens* (see MD plate 211), grows under mountain conifers in the spring and early summer; *Otidea* species are earlike and stand erect. See MD 831–838 and plates 208–209.

Blue stain (*Chlorociboria aeruginascens*) is a tiny blue-green to turquoise cup fungus that stains wood blue-green. The wood was once used to make inlaid objects.

Domestic Cup Fungus
(*Peziza domiciliana*)

Other Names: Domicile Cup Fungus.

Key Features:
1. Mushroom shaped like a shallow cup when young, becoming flat or wavy with age.
2. Upper surface off-white to tan or brown, bald.
3. Stalk short or absent (most evident when young).
4. Growing indoors.

Other Features: Small to medium-sized; flesh rather fragile; underside same color as upper surface or paler.

Where: Alone or in clusters in bathrooms, cellars, greenhouses, closets, and cars, under porches, behind refrigerators, etc. It grows on many substrates, including plaster, cement, wood, and carpets; common and widespread.

Edibility: Not recommended.

Note: This is one mushroom you don't have to stray far from home to find! Many similar brown cup fungi grow on dung, rotten wood, charcoal, and forest humus. See MD 818–825 for more information.

Crown Cup (*Sarcosphaera crassa*)

Other Names: Crown Fungus, Pink Crown, *Sarcosphaera eximia*, *S. coronaria*.

Key Features:
1. Mushroom beginning as a hollow ball, then splitting at the top to form a crown with several rays.
2. "Crown" typically 2–5" broad when expanded.
3. Upper surface (interior) white or grayish at first, pinkish to purplish, purple-brown, or tan after splitting open.
4. Underside (exterior) whitish, *without* brown hairs.

Other Features: Young specimens usually developing underground; interior bald; exterior usually dirty; flesh whitish and brittle; stalk short or, occasionally, absent.

Where: Solitary or in small clusters in soil (usually half-buried) under conifers; widespread and common, especially in mountains. It is most abundant in the spring during morel season but can be found at other times.

Edibility: Not recommended. One authority compares the texture to "a rubber eraser softened by time." It is difficult to clean and has been known to cause digestive upsets. Furthermore, it concentrates significant amounts of arsenic from the soil.

Note: This distinctive fungus is often brought home by skunked morel hunters who wonder whether it is edible. It varies considerably in color but usually shows at least a tinge of purple or pink. Dead man's hand (*Scleroderma geaster*) can be crownlike, but has a solid interior when young that turns into spore powder after it splits open. See p. 240 for another photo, and MD 825–826.

Oregon White Truffle
(*Tuber gibbosum*)

Key Features:
1. Mushroom potato-like, buff to tan, brown, or red-brown (darker when mature).
2. Exterior *without* warts.
3. Interior solid, very firm, marbled with veins.
4. Interior white when young, dark brown or reddish marbled with whitish veins when mature.
5. Stalk absent.
6. Odor at maturity strong and somewhat garlicky.
7. Growing underground in association with Douglas-fir.

Other Features: Maturing very slowly (usually several weeks); typically ½–2" broad (occasionally larger); exterior often cracked at maturity; skin thin.

Where: Buried in the ground near or under Douglas-fir in forests or on Christmas tree farms; West Coast, especially common in Oregon. Several can often be found in the same area by gently raking the soil.

Edibility: Some chefs rate it on a par with the famous white truffle of Italy.

Note: Truffles are small, difficult-to-identify, underground mushrooms whose spores are disseminated by rodents and other foraging mammals that sniff them out. Small excavations in the forest humus are often a sign that truffles or truffle-like mushrooms are in the vicinity. This truffle is best recognized by its habitat, and when fully mature, by its odor. See MD 841–844, 854–855, 858–859 and plates 212–213.

Lobster Mushroom
(*Hypomyces lactifluorum*)

Key Features:
1. Mushroom engulfing its host (a gilled mushroom) in a layer of bright orange to red or sometimes purple-red tissue, making the whole thing look more or less like an upside-down pyramid.
2. Gills of the host mushroom reduced to blunt ridges or not visible at all, the surface covered instead with tiny pimples (use hand lens!).
3. "Gills," when present, *not* forked or veined.
4. Flesh white and crisp when fresh.

Other Features: Medium-sized; cap sometimes yellower than underside; flesh softer in old age; odor mild.

Where: On ground in woods, where its favorite host, the short-stemmed russula (*Russula brevipes*) grows; widespread, especially common under conifers in the Pacific Northwest and Southwest.

Edibility: Delicious when still crisp — a definite improvement over the short-stemmed russula! It should be sauteed.

Note: This fluorescent-looking mushroom parasite is sometimes mistaken for a chanterelle, but is oranger or redder with unforked "gills" (if any) and a finely roughened or pimpled underside. It belongs to a group called the flask fungi; other flask fungi are shown on the facing page. See MD 884 and plate 216.

Hypomyces luteovirens covers at least the gills of milk caps and russulas in bright green or yellow-green tissue; so far as we know it is edible.

My Most Memorable Mushroom Hunt

It occurred during one of the worst periods of my life. In less than three months my family and I were hit with an incredible series of calamities: serious illness, death, a miscarriage, the birth of a baby with congenital defects. For days the pain was so intense I was sure it would never go away. At other times, I felt numb.

Then a friend called, asking if we could get together to hunt some mushrooms. There weren't any at the first place we went, and we had a flat tire on the way to the second. But after that it was perfect — one of those cool, hushed, misty days.

The fruiting was meager: lactarius, collybias, polypores, a few old chanterelles, and lots of russulas. But there were creeks to cross and mists dancing in the air, and plenty to smell and taste and look at and ponder and chuckle about. It wasn't the mushrooms that made the day so special, but the sharing of wonder with someone who loved them as much as I did. I wanted the day to go on and on, and I knew that night that I was going to be all right.

—*Theresa Ray*

This whitish species, *Hypomyces lateritius*, can be eaten when it grows on green-staining milk caps (see pp. 15–17).

Corn Smut (*Ustilago maydis*)

Other Names: Cuitlacoche, Huitlacoche.

Key Features:
1. Growing on corn as a clump of tumorlike growths that look like overgrown corn kernels.
2. Growths whitish when very young but soon becoming gray and shiny.
3. Growths at first firm inside, then becoming black and grainy.

Other Features: Individual growths up to ¾" long, rupturing at maturity; stalk absent.

Where: On corn in fields and gardens; widespread and common.

Edibility: Excellent; it is considered a great delicacy in Mexico. It should be eaten when gray on the outside, black and juicy or slightly grainy within. Diana Kennedy (see facing page) gives several recipes for it. (For one of the most delicious soups in the world, try her *sopa verde de elote*, which uses corn and tomatillos but not corn smut.)

Note: Also known as "porn on the cob," corn smut is not a mushroom in the usual sense, but it is definitely a fungus. Mexican farmers welcome its appearance in their cornfields, because it fetches a higher price than the corn! For more information and cooking suggestions, see the following two books by Diana Kennedy: *Mexican Regional Cooking* (New York: Harper Colophon Books, 1984); and *The Art of Mexican Cooking* (New York: Bantam, 1989).

Cuitlacoche Filling

(for quesadillas or crepes)

Cut the fungus and corn kernels from the cob, keeping them as whole as possible to preserve their texture. Roughly chop the fungus and weigh out 1½ pounds (about 6 cups). Also char and peel 4 *chiles poblanos* (a medium-spicy chile), remove the seeds and slit vertically into ½" strips.

Heat 3 T. safflower oil in a frying pan, add 2 T. finely chopped white onion and 2 small garlic cloves, peeled and finely chopped. Fry until translucent — about 3 minutes. Add the chiles and fry for another minute. Then add the *cuitlacoche* and salt to taste, cover the pan, and cook over medium heat, shaking occasionally, for about 15 minutes. The smut should be tender and moist but not soft and mushy. (If too dry, sprinkle on ¼ cup water before covering; if too juicy, remove lid to drive off excess moisture.) Finally, stir in 2 T. chopped *epazote* (Mexican-tea) leaves and cook, uncovered, for another 2 minutes.

—adapted from *The Art of Mexican Cooking*
by Diana Kennedy

Toward a Mushroom Hunting Ethic

What are the impacts of gathering wild mushrooms? Since mushrooms are the "fruit" of inconspicuous fungal mycelia that often live for many years, picking them has been compared to removing berries from a bush: no harm is done so long as the environment that supports the fungus is not disrupted and some mushrooms are left to shed their spores. (Many actually shed spores *before* being picked.)

While this analogy is a fairly good one, conditions have changed: mushroom hunting has become a popular pastime, and there has been a dramatic increase in the commercial harvest of certain species, primarily for export to Japan (matsutake) and Europe (chanterelles and morels). The matsutake harvest is of special concern because of the high prices paid for unopened "buttons" that have yet to form spores, and the manner in which some pickers rake or otherwise disturb the humus or moss in their search for the buried buttons.

Some concerned individuals have invoked the spectre of species extinction, believing that commercialization inevitably results in the degradation or "rape" of the landscape. But comparing mushrooms to trees, buffalo, or whales (as alarmists often do) is bad biology. Loggers don't harvest trees from the same stump year after year. When you kill a whale, you are eliminating that organism and all its potential progeny from the environment. But when you pick a mature wild mushroom, you may actually aid spore dispersal by toting it around; and if the plucked mushroom hasn't yet shed spores, at least the organism that produced it is still there, capable of producing more mushrooms in the future.

Ongoing studies bear this out. They show that picking all the mushrooms has no adverse impact on future crops in that area, and may even have a slight stimulating effect. However, there are no studies that measure the ability of a species' spores to colonize new locations. Established mycelia, or "mushroom patches," that continue to produce mushrooms for many years could actually mask the reduced recruitment that overharvesting might cause. There are other concerns as well, for instance, the impact of intensive harvest on genetic diversity, and on the forest food web. But often missing is a realistic perspective on the most significant danger to mushrooms and virtually everything else alive: destruction of habitat. Instead, anger is directed at commercial mushroom pickers, most of whom are poor, unorganized, and much more visible and vulnerable than some of the larger and more powerful entities that degrade habitat.

Obviously, it is impossible to determine the long-term impacts of mushroom hunting in the short term. But we can take some comfort from the fact that intensive mushroom harvest is *not* new. It *feels* new to us because we live in a fungophobic culture that until recently did not pay much attention to mushrooms. But there is a very long and colorful history of intensive mushroom gathering around the world: from Italy to Russia, Siberia, China, Japan, Thailand, Kashmir, Turkey, Mexico, Chile, Morocco, Zimbabwe, etc., edible mushrooms are still plentiful where there is healthy habitat to support them. (As I write this, the price for dried boletes has been depressed by a bumper crop from eastern Europe.) This suggests that there is no reason to panic, and that if long-term studies do reveal adverse impacts, it will be possible to soften or eliminate them by modifying collecting practices once the effects are known.

There are those who advocate a more "conservative" approach: that of prohibiting or severely restricting mushroom picking until it can be "proven" harmless in the long term. But what is deemed conservative by these people is perceived as radical by those whom it affects. In the absence of demonstrated harm, such restrictions exact a high social cost because they seem arbitrary and mean-spirited. In other words, the largely urban-suburban vision of nature as a beautiful, peaceful refuge from the stresses and conflicts of civilization, is in fundamental conflict with the rural or less "civilized" perception of nature as a provider of sustenance and wealth. Therein lies a great irony: it is the cities that suck food, energy, and resources from the landscape, yet there is a long and tragic history of industrial and agricultural peoples persecuting "savage" outsiders (in the most literal sense of the word) who hunt and gather.

Increased competition for mushrooms is another source of friction. Although this is a territorial issue, some amateur pickers and mushroom clubs have tried to frame it as a moral one, claiming that they love mushrooms while commercial pickers only love money. But in my experience, *both* groups of people love mushrooms, and *everybody* loves — or at least needs — money. In fact, I am more impressed by similarities between the two groups. There *are* differences, particularly in economic class, but both groups share a passion for mushrooms and an impressive knowledge of the natural world; both are fascinated by the challenge of finding mushrooms and "figuring them out"; and both groups have an obvious stake in the sustainability of what they do. Rather than spending precious energy complaining *about* commercial pickers, noncommercial mushroom hunters have more to gain by exchanging knowledge and information *with* them.

Because many land managers and public officials are unable or unwilling to distinguish between commercial and noncommercial pickers, branding commercial pickers as "money grubbers" and "rapists" has tended to tar *all* mushroom hunters. The result has been a flood of ill-conceived regulations that severely restrict or ban the gathering of mushrooms on public lands, even for personal use. In much of California mushroom hunting is now virtually illegal, and an amateur picker from Oregon reflects a growing sentiment when she says: "I feel much more oppressed by government overregulation than by commercial pickers."

Land managers and rangers can be *very* territorial, and some seem to regard the plucking of a mushroom as an environmental sin on a par with felling a tree or shooting a pelican. No doubt the admonition to "take only photographs, leave only footprints," is desirable in certain sensitive or pristine areas. But it is inappropriate in most situations because it is a fundamental denial of who we are: creatures of the earth whose most ancient heritage (or birthright, if you will) is foraging for food in the forest.

This enforced separation of human being from the natural world can only undermine support for the public acquisition and protection of habitat. The more we view the forests and fields as a *resource* — one that yields, among other gifts, a beautiful, delicious, valuable, and renewable mushroom crop — the more likely we are to cherish that resource. Protecting habitat for aesthetic reasons alone lends credence to the argument that a few widely scattered parks are enough. But harvesting a sustained yield of mushrooms, fish, and other wildlife creates a *quantitative* demand for habitat on top of a qualitative one. For this reason I believe we should strongly encourage foraging for mushrooms. It is a relatively benign use of our forests that can foster a powerful love and *need* for the natural world, and increase our stake in protecting and enhancing it.

Western North America has a rich and diverse mushroom bounty that can be gathered and enjoyed by thousands. Obviously, it should be managed in a way that ensures sustained yield and a healthy forest ecosystem. But an enlightened policy requires a wealth of data that can only be accumulated with research, over time. What do we do in the meantime?

First, we should acknowledge the value and desirability of mushroom hunting by allowing it on most public lands, whether for research, pleasure, or profit. There should be some parks where picking is not allowed; there should also be some public lands that allow limited gathering for personal use but are off-limits to commercial collecting.

Second, mushroom hunters and non-mushroom hunters alike should refrain from trying to restrict the activities of others for ideological reasons alone, or without clear evidence that what others are doing is harmful.

And third, as individuals we should be gentle with the forest. Don't leave an unsightly trail of trash behind, but do leave some mushrooms for those who follow in your footsteps. You may be one of them!

Checklist and Index to Common Names

A checklist box precedes the primary common name of each mushroom that is fully described; some other widely used names are also listed but not boxed.

257

Index to Scientific Names

"The mushroom-scented air of the birch groves is far dearer than the fragrance of the magnolia." — Paustovskiy

Photo Credits and Acknowledgments

All photographs are by David Arora except for the following: Catherine Ardrey pp. 24, 42, 135 right, 149 top, 178 top left and bottom, 179 bottom, and 235 bottom left; Harley Barnhart p. 93 top right, 218; Alan Bessette p. 93 top left; Mike Bonnicksen/Wenatchee World p. 231 bottom; Frank Dutra p. 83 top; Bill Everson p. 103 bottom; Tarmo Hannula p. *iv*; Rick Kerrigan p. 116; Taylor Lockwood pp. 8 top right, 10, 46, 72 left, 214 bottom, 215 bottom left; Bill McGuire p. 103 top; Grant Neville p. 145 bottom; Phillip Orpurt p. 93 bottom, Herb Saylor p. 199; Kit Scates pp. xx, 60, 72 right, 192, 234, 236, 240 top left; Joy Spurr p. 167; Paul Stamets pp. 124 top, 130; Larry Stickney p. 94 bottom; Jim Trappe p. 247 (all); Bob Winter pp. 34 top left, 39 right; and Greg Wright pp. 37 left, 227 right.

The author would like to express his deep gratitude to the many people who helped create this book, including the photographers listed above, all those who contributed stories, anecdotes, and recipes, and the following people for their largely unseen but crucial contributions: Ed Aguilar, Joe Ammirati, Scott Anderson, Jim Arnold, Harbans and Shirley Arora, Brenton Beck, Anne Canright, Jane Christmann, Frank Dutra, Mary Engbring, John Feci, Gregg Ferguson, David Grimes, Katie Hart-Thelander, Laurel Herter, Bo Hinrichs, Grainger Hunt, Diana Kennedy, John Lane, Shannon Loch, Lia Matera, Judith Scott Mattoon, Charlie McDowell, Luen Miller, Charles Peck, Enrico Re, Miriam Rice, Clark Rogerson, William Rubel, Maya Szpakowski, Paul Stamets, Carl Thelander, George Tsapanos, Andrew Weil, and Brigid Weiler.

QUICK KEY TO MUSHROOMS WITHOUT GILLS

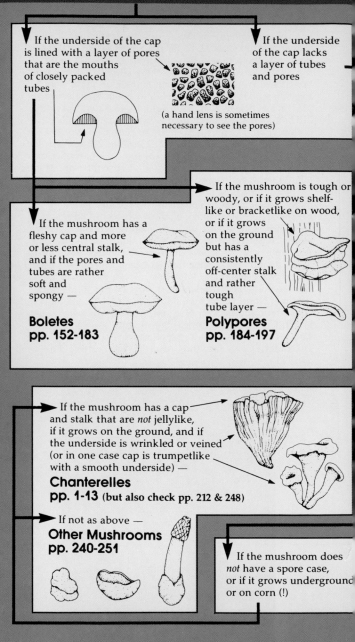

If the underside of the cap is lined with a layer of pores that are the mouths of closely packed tubes

If the underside of the cap lacks a layer of tubes and pores

(a hand lens is sometimes necessary to see the pores)

If the mushroom has a fleshy cap and more or less central stalk, and if the pores and tubes are rather soft and spongy —

Boletes
pp. 152-183

If the mushroom is tough or woody, or if it grows shelf-like or bracketlike on wood, or if it grows on the ground but has a consistently off-center stalk and rather tough tube layer —

Polypores
pp. 184-197

If the mushroom has a cap and stalk that are *not* jellylike, if it grows on the ground, and if the underside is wrinkled or veined (or in one case cap is trumpetlike with a smooth underside) —

Chanterelles
pp. 1-13 (but also check pp. 212 & 248)

If not as above —
Other Mushrooms
pp. 240-251

If the mushroom does *not* have a spore case, or if it grows underground or on corn (!)